MOHAMMED BENSLIMANE

Phyto-purification

MOHAMMED BENSLIMANE

Phyto-purification

Concept, process and experimental trials

ScienciaScripts

Imprint
Any brand names and product names mentioned in this book are subject to trademark, brand or patent protection and are trademarks or registered trademarks of their respective holders. The use of brand names, product names, common names, trade names, product descriptions etc. even without a particular marking in this work is in no way to be construed to mean that such names may be regarded as unrestricted in respect of trademark and brand protection legislation and could thus be used by anyone.

Cover image: www.ingimage.com

This book is a translation from the original published under ISBN 978-620-6-72335-6.

Publisher:
Sciencia Scripts
is a trademark of
Dodo Books Indian Ocean Ltd. and OmniScriptum S.R.L publishing group

120 High Road, East Finchley, London, N2 9ED, United Kingdom
Str. Armeneasca 28/1, office 1, Chisinau MD-2012, Republic of Moldova, Europe
Printed at: see last page
ISBN: 978-620-8-24033-2

Contents

Mohammed BENSLIMANE

Lecturer-researcher Fac. SNV- Geomatics, ecology and environment laboratory. Universite de Mascara, B.P n°305, Mascara, Algerie. Email : med_benslimane@yahoo.fr

Phytodepuration, as a wastewater treatment process for small agglomerations with a low pollution load, has amply demonstrated its performance in many countries around the world. In Algeria, few attempts have been made in terms of investment, preferring to develop research on experimental pilots. Our laboratory has carried out two such studies: a reduced-scale model set up at the University of Mascara's experimental farm, and a full-scale pilot at Brezina (El Bayadh). The TMEs are stored in a finishing tank before being recycled for crop irrigation. Because of its significant and encouraging results, the phytodepuration process offers a completely justifiable opportunity to contribute to the challenge of water saving and efficient water management, particularly in small communities. The aim of this book is to shed some light on this ecological purification process, followed by an example of a sizing calculation.

INTRODUCTION

Aware of the importance of protecting the environment as part of sustainable development, a great deal of effort has been put into wastewater treatment in recent years. However, the techniques used are still far from meeting all the demand.

As things stand, this ambiguity poses a problem of pollution and public health, since experience shows that the intensive technologies initially developed for use in urban areas are not necessarily suited to semi-urban and rural areas, which account for almost 40% of the population.

The disadvantages raised by managers of small local authorities are several-fold (high investment costs, lack of technological expertise, local management difficulties, etc.).

In contrast, the revolution in wastewater treatment regulations in terms of standards for discharging wastewater into the natural environment reminds us of the need for ***appropriate treatment*** of such discharges for all polluting activities, in order to ensure both the immediate environment and the protection of groundwater resources and natural watercourses.

The expected objectives of wastewater treatment also include a substantial increase in the resource and its recycling in the economic water circuit, particularly for the needs of crops less agriculture.

Faced with the combined challenges of the spectre of a water shortage exacerbated by climate change on the one hand, and the ever-increasing state of environmental degradation on the other, the use of ecological wastewater treatment systems is becoming increasingly popular because of their purification performance, their advantages in terms of operation and maintenance, and their lower investment costs for small urban and rural communities.

Among these ecological systems, ***phyto-purification***, or plant filters made from macrophyte and microphyte plants, is currently attracting growing interest. One of the reasons for this is the way it can be installed and the means available for setting up individual and/or collective autonomous wastewater treatment systems.

This book is intended as a contribution to knowledge of this process, which has proved effective in many countries because of its simplicity. In the Algerian case, the constraints of space and sunshine do not arise, hence the interest in developing it in small towns with a low pollution load.

The structure of the book is divided into two complementary parts, one dealing with laboratory tests using highly instructive teaching aids intended for graduate students, the other more spatial, carried out on a real scale for professional use.

The aim of this work is to demonstrate this phytodepuration system, with the help of experimental pilots installed at the SNV Faculty of the University of Mascara, using different ways of circulating wastewater through a series of planted basins. The method involves using local plant species (reeds, typha, rushes, irises, duckweed) planted in filters, with the wastewater passing through a series of basins.

Purification performance is assessed using physico-chemical and bacteriological analyses at the inlet and outlet of the experimental pilot plant. For information

purposes, the results are also compared with those of the El Keurt lagooning plant.
The principle adopted for the Brezina experimental pilot plant is based on the construction of two treatment processes. The first (A) consists of a horizontal flow of wastewater through three ponds in series. The second (B) has a horizontal sub-surface flow. The two processes meet in a buffer tank, which is used for storage prior to reuse in a 5-hectare arboretum of forage species.
The pilot project, carried out as part of an Algerian-Italian partnership to combat desertification, is limited to treating a fraction (30%) of the wastewater discharged by the oasis settlement of Brezina (El Bayadh). The treatment process is equipped with a diversion of the town's waste water, a primary sedimentation tank (*Imhoff* type*) and* a screening machine to eliminate solid matter.
A 0200 mm manhole allows gravity flow to the gravel and sand filters and phragmites (common reed) plants. Finally, there is a valve for regulating and measuring the flow through each of the treatment beds. It should be noted that access is provided at all the water distribution pipe connection points to ensure that the facilities can be maintained.
Thus, the phyto-purification experiments carried out both on a reduced scale and on a natural scale present an effective alternative for the treatment of waste water for the benefit of local communities, which needs to be developed and generalised, particularly in mountain and oasis ecosystems.

Chapter I

GENERAL INFORMATION ON PURIFICATION

Introduction

Wastewater treatment has always been considered from the point of view of reducing a pollutant load, whether organic, particulate or linked to the presence of pathogenic germs.

One of the most important aspects of such a process is undoubtedly the effectiveness of the treatment that characterises the wastewater, despite the considerable complexity of the influencing physical, chemical, biological and mechanical parameters.

The choice of treatment methods depends on a number of factors, the most important of which are the quality of the effluent, the size of the plant and the quality requirements of the wastewater in terms of its reuse in the economic water cycle.

1.1 Definition of waste water

Wastewater is all water from domestic, agricultural and industrial activities that is laden with toxic substances and finds its way into sewage pipes. Wastewater also includes rainwater and its pollutant load, which causes all kinds of pollution and nuisance in the receiving environment [1].

1.2 Sources of waste water

Urban wastewater is made up of household wastewater (washing the body and clothes, washing premises, kitchen wastewater) and black water containing faeces and urine. All this effluent is more or less diluted by road washing water and rainwater [2].

Depending on the case, water from industrial and agricultural sources may also be included. Water collected in this way in a sewer system appears as a cloudy, generally greyish liquid, containing suspended matter of mineral and organic origin in extremely variable concentrations. According to [3], used water is water discharged after industrial, domestic or agricultural use.

1.2.1 Domestic origin

This is water polluted by all domestic activities, i.e. at home. We can distinguish :

- Grey water: this is water from showers and kitchens. It is generally loaded with grease, surfactants (soaps, washing powder), solvents, food residues, etc.
- Black water: this is toilet water. It is made up of faeces, urine and paper [4].

1.2.2 Industrial origin

Industrial waste and effluents largely determine the quality and pollution rate of this wastewater. Industrial establishments use a large quantity of water which, while necessary for their proper operation, is only consumed to a very small extent, the remainder being discharged. The main industrial discharges have been classified according to the nature of the harm they cause:

- Pollution due to suspended mineral matter (coal washing, quarrying, sand and gravel sieving, phosphate fertiliser industries....);
- Pollution from mineral solutions (pickling and galvanising plants, etc.);
- Pollution from organic matter and fats (food processing industries, meat processing, paper pulp, etc.);
- Pollution from various hydrocarbons and chemicals (oil refineries, pig farms, pharmaceutical products);
- Pollution from toxic discharges (untreated radioactive waste, radioactive effluents from nuclear industries....).

Industrial wastewater generally has a more specific composition, directly related to the type of industry in question. Irrespective of the load of organic or mineral pollution, and whether or not it is putrescible, it may have its own toxicity characteristics linked to the chemicals transported [2].

1.2.3 Agricultural origin

This is water used for agricultural purposes. This is not the first water that comes to mind when we think of sanitation, even though it represents a large majority: in fact, 70% of water drawn from the natural environment is consumed for agricultural purposes according to the Food and Agriculture Organization of the United Nations. For example

- White dairy water
- Manure water
- Drainage water, etc. [4].

1.2.4 Rainfall origin

Rainwater contains impurities. It can be polluted by air pollution (e.g. acid rain). What's more, once it hits roofs and the ground, it runs off and takes with it whatever it comes across. In this way, they can degrade the quality of watercourses.

1.3 Characterisation of waste water

In this section we will outline the key physico-chemical parameters analysed during the experimental phase, as well as the bacteriological parameters most commonly found in wastewater.

I.3.1 Physical parameters

a) Colour: Fresh wastewater is normally brown and yellowish, but over time it turns black.

b) Suspended matter: this is insoluble solid matter suspended in a liquid and visible to the naked eye.

c) Temperature: For waste water, this is corrected to the outside temperature, but is warmer because almost nobody takes a cold shower.

d) Turbidity: Due to the suspended matter, waste water will have a higher turbidity.

e) Hydrogen potential (pH)

The acidity, neutrality or alkalinity of an aqueous solution can be expressed by the concentration of H3O + (noted H+ for simplicity). To make this expression easier, we use the decimal logarithm of the inverse of the H+ ion concentration: this is the pH [5].

$$pH = \log 1/ [H +]$$

f) Conductivity

This is the property of water to encourage the passage of an electric current. It is due to the presence in the medium of ions that are mobile in an electric field. It depends on the nature of these dissolved ions and their concentrations [6]. The electrical conductivity of water is the conductance of a column of water between two 1 cm metal electrodes^2 . The unit of conductivity is the siemens per metre (S/m).

$$1 \text{ S/m} = 104 \text{ pS/cm} = 103 \text{ mS/m } [7]$$

g) Chemical Oxygen Demand (COD)

This is a measure of the amount of organic matter in wastewater as a function of the oxygen required to oxidise it (1H2O3).

h) Biochemical oxygen demand (BOD)

In practical terms, biochemical oxygen demand should make it possible to assess the putrescible substance load of the environment in question, its self-cleansing capacity and to deduce the maximum acceptable load, mainly at the primary treatment stage of wastewater treatment plants [2].

The biochemical oxygen demand after 5 days (BOD_5) of a sample is the quantity of oxygen consumed by the aerobic microorganisms present in this sample for the biochemical oxidation of organic and/or inorganic compounds [7].

i) Nitrates

Nitrates occur naturally in water, largely as a result of run-off from the soil that makes up the catchment area. Their natural concentrations do not exceed 3 mg/l in surface water and a few mg/L in groundwater.

The nature of drainage areas therefore plays a key role in their presence, and human activity accelerates the process of enriching water with nitrates [7].

Nitrate levels have been rising in recent years, by between 0.5 and 1 mg/l/year, and even 2 mg/l/year in some regions. There are several reasons for this increase:

- *Agriculture*: intensive farming with massive use of nitrogen fertilisers and discharges of livestock effluent. This source accounts for 2/3 of nitrate inputs into the natural environment;
- *Urban*: discharge of waste water from waste water treatment plants where the elimination of nitrogen is not total and which may discharge nitrates or ammonium ions which will be transformed into nitrates in the natural environment. This source represents 2/9 of inputs;
- *Industrial*: waste from mining industries, in particular nitrogen fertiliser manufacture. This source accounts for 1/9 of inputs [6].

j) Nitrogen

Nitrogen in water can be organic or mineral. Organic nitrogen is mainly made up of compounds such as proteins, polypeptides, amino acids and urea. In most cases, these products are only present in very low concentrations. Mineral nitrogen (ammonia, nitrate, nitrite) accounts for the majority of total nitrogen [2].

k) Phosphorus

Phosphorus is naturally present in very small quantities in the soil and water. A high concentration of phosphates can lead to the proliferation of algae, which is a nutrient for plants. Algae are responsible for the eutrophication of stagnant water.

However, phosphorus is the limiting factor that can be effectively controlled to reduce eutrophication.

According to one study, 1g of phosphate-phosphorus (PO_4-P) can cause an algae bloom of 100g. When these algae die, they require around 150g of oxygen to decompose. This phenomenon begins at very low P-PO4 concentrations:

- 0.1-0.2 mg/l in running water
- 0.005-0.01 mg/l in stagnant water.

The chemical forms of phosphorus in wastewater vary greatly. They can be soluble or particulate, mineral or organic.

Total Phosphorus = Particulate Phosphorus + Dissolved Phosphorus = Mineral Phosphorus + Organic Phosphorus

In order to comply with the recommended thresholds, a target phosphorus elimination process is required in wastewater treatment plants. This process can be chemical or biological, and often both (H_2O_3).

l) Sulphate

The sulphate ion concentration of natural water varies greatly. In soils that do not contain a significant proportion of mineral sulphates, it can reach 30 to 50 mg/L, but this figure can be greatly exceeded (up to 300 mg/L) in areas containing gypsum or when the contact time with the rock is high.

The sulphate content of the water must be related to the alkaline and alkaline-earth elements of the mineralisation. Their presence in water is generally due to discharges from bleaching plants (wool, silk, etc.), cellulose manufacturing plants (paper pulp, etc.) and chlorination units.
The reducing properties of sulphites are also used in boiler water to prevent corrosion caused by the presence of dissolved oxygen; injection into the circuit is usually continuous at a concentration of 20 mg/l.
However, an excess of sulphite ions in boiler water can have harmful effects because it lowers the pH and can then promote corrosion. When released into the environment, sulphites combine with oxygen to form sulphates [2].

I.3.2 Bacteriological parameters

Bacteria are ubiquitous in nature, as they were probably the first living beings to appear on earth (archeobacteria). Only a few dozen species are adapted to humans: most are harmless or even useful, being commensal and part of the cutaneous, digestive, oral and genital flora; some are pathogenic, opportunistic; a minority are regularly pathogenic [2].
The main bacterial indicator for monitoring faecal pollution is the coliform group of pathogenic microorganisms, e.g. Salmonella, Giardia lamblia, Escherichia col...etc [8].

a) Faecal coliforms

Fecal coliforms, or thermotolerant coliforms, are a subgroup of total coliforms capable of fermenting lactose at a temperature of 44.5°C. The species most frequently associated with this bacterial group is *Escherichia coli* (*E. coli*) and, to a lesser extent, certain species of the genera *Citrobacter*, *Enterobacter* and *Klebsiella* [9].
However, *E. coli* represents 80-90% of the thermotolerant coliforms detected [10]. Although the presence of faecal coliforms usually indicates contamination of faecal origin, several faecal coliforms are not of faecal origin, but rather come from water enriched with organic matter, such as industrial effluents from the pulp and paper or food processing sectors [11] (Fig.1).

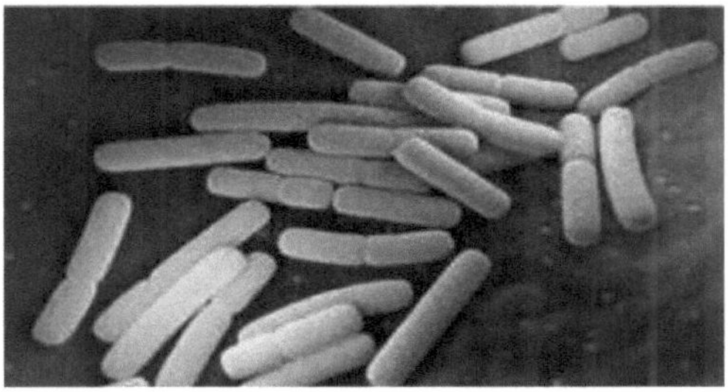

Figure 1 - Microscopic view of faecal coliforms.

b) Enterococci

Enterococci are Gram-positive bacteria in the form of diplococci or chain shells. They are facultative anaerobes, immobile and lack a capsule.
These germs are undemanding and can survive in hostile conditions, such as highly alkaline or salt-rich environments (e.g. bile salts), or extreme temperatures (10°C to 60°C). They can also survive for long periods on all kinds of surfaces [12].
In water, they are indicators of faecal contamination, such as colibacilli (Fig.2).

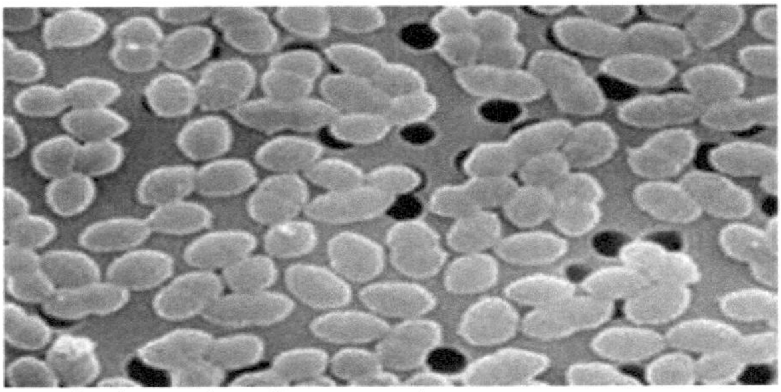

Figure 2 - Microscopic observation of enterococci

c) Sulphate-reducing bacteria

Sulphate-reducing bacteria are anaerobic bacteria, such as Desulfovibrio; it is the sulphates that replace oxygen for cellular respiration. During metabolism, sulphates are reduced to sulphides.

These sulphides can cause corrosion, known as bio-corrosion, particularly on sheet piling and hydrocarbon tanks [7]. (Fig.3)

$$SO_3^{\cdot 2} + 6H^{+}\ 6 - \wedge\ S_{2-} + 3H_2O$$

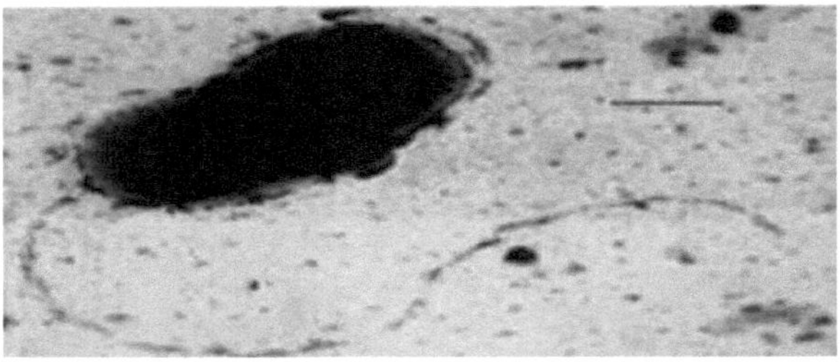

Figure 3 - Microscopic view of a sulphate-reducing micro-organism.

1.4 Waste water treatment

Water purification is a set of techniques that involve purifying water, either to reuse or recycle wastewater in the natural environment, or to transform natural water into drinking water.

The waste water treatment sector recommends various techniques at different technological levels, often very sophisticated. These are illustrated as being classic treatment methods, as well as new techniques aimed at protecting the environment and safeguarding the natural environment, such as lagooning or phytodepuration.

1.4.1 Definition of a wastewater treatment plant

This is an installation designed to treat domestic or industrial wastewater and rainwater before it is discharged into the natural environment.

The aim of treatment is to separate the water from substances that are undesirable for the receiving environment. The first attempt to treat wastewater was invented in 1914 by British scientists [13].

1.4.2 Objectives and conditions (siting of a WWTP)

- Protecting groundwater from pollution.
- Preventing farmers from irrigating farmland with wastewater.

- Minimising the risk of water-borne diseases.
- Significant water savings.
- Avoid overexploitation of groundwater [13].

The wastewater treatment plant must be built in accordance with the following conditions:

- Avoid flood-prone areas as far as possible.
- Consideration of urbanised and potentially urbanised areas due to unpleasant odours, airborne diseases, etc.
- It should be located downstream of the sewerage system (at the outlet) to avoid the need for lifting and the consequent high cost.
- The wastewater treatment plant should be located at a point where wastewater collection is at its highest (e.g. a strategic point between two built-up areas).
- The discharge of waste water must not affect the receiving environment [13].

1.4.3 Risks associated with the wastewater treatment plant

Knowing full well that there is no such thing as zero risk, it is highly likely that there will be some risks at a wastewater treatment plant, in particular :

- Risk of personal falls in the immediate vicinity of equipment and materials.
- The risk of fire due to the presence of fermentation gases or residues of flammable products, necessary or accidentally introduced by waste water.
- The risk of respiratory contamination.
- Electrical risks depend largely on the materials used for the various installations and the design of these installations.
- Chemical risks arising from physico-chemical water treatment processes, sludge conditioning and effluent disinfection techniques
- Infectious risks due to the presence of a microbial population in the wastewater, which contaminates the public in the plant [13].

► **Main pathogenic germs encountered and the disorders they cause** [14].

Table 1 - Bacteria likely to be transmitted by wastewater.

Illness agent	**Reservoir**	**Transmission mode**
Escherichia coli Gastroenteritis	Men Pets	Digestive tract (manuportee)
Salmonella sp Typhoid fever and salmonellosis	Humans (healthy carriers) Infected cattle - infected animals - birds	Digestive tract
Shigella sp Bacillary gastroenteritis (dysentery)	Men Pets	Digestive tract
Brucella sp Brucellosis	Infected cattle	Direct contact with water

Table 2-Viruses likely to be transmitted by waste water.

Agent **Disease**	**Reservoir**	**Transmission mode**
Enterovirus (sp poliovirus) Gastroenteritis, heart problems...	Men Small animals	Digestive tract
Poliovirus Poliomyelitis	Men	Digestive tract
Hepatitis A and E viruses	Man and primates	Digestive tract

Hepatitis		
Rotavirus gastroenteritis	Man and domestic animals	Digestive tract (faecal/oral)
Adenovirus Respiratory diseases	Man and other primates	Airborne (direct contact)

Table 3- Parasites likely to be transmitted by wastewater.

Agent Diseases	Reservoir	Transmission mode
Protozoa Entamaba histolycita Amoebiasis Giardia lambilia Giardiosis	People, livestock, pets	Digestive tract
Helminths (round) Ascaris duodenale Ascaris lumbricoides Ascariasis Ankylostoma duodenale Hookworm Strongyloides vermicularis Strongylosis Anguillulosis	Men Men Man, dog	Digestive tract
Helminths (flatworms) Taenia saginata Tenia Fasciola hepatica Great fluke	Men Cattle	Digestive tract

1.5 Waste water treatment

Domestic wastewater can be treated by various processes based on physical, chemical and biological phenomena, with varying degrees of purification and costs depending on the level of quality required by the receiving environment.

1.5.1 Stages in the treatment process

a) Pre-treatment (physical)

The aim is to eliminate the coarsest solid elements or particles that could cause further treatment or damage the equipment. These pre-treatment processes are: screening, grit removal, oil removal and degreasing [15].

♦ Degritting

Screening, the first stage in the treatment of raw effluent, is used to retain bulky waste (floatables, filings, etc.) through screens with different mesh sizes. Depending on the distance between the bars of the screen, operations such as pre-screening, medium screening, fine screening and sieving can be distinguished. If the bars are 30 to 100 mm apart, this is known as pre-screening.

This is followed by medium screening with bars 10 to 30 mm apart, and finally fine screening with bars less than 10 mm apart. The screens can be vertical, inclined at 60° to 80° to the horizontal for greater efficiency, or curved.

Fine screens can be stepped, or have a rotating drum, a spiral screen or a filter chain. Average effluent velocities vary between 0.60 and 1.40 m/s at peak flow. Cleaning is carried out manually from upstream using a rake placed on the screen by a technician (Fig. 4).

Figure 4 - Degreaser.

In rural areas, it would seem prudent to provide a grille with widely spaced bars. The spacing should be between 40 and 50 mm, and a bypass with a guard grille with even wider bars should be provided to compensate for a lack of surveillance [16].

♦ **Sand, grease and oil removal**:

Grit removal consists of extracting the heavy elements from the raw effluent, such as quick settling tables, sand, gravel and other mineral particles, in sizes ranging from 100 to 200 [17].

Degreasing is the removal of non-soluble fatty particles (10 to 20% of fats) by flotation. Oil removal consists of eliminating the oils present in industrial wastewater by manual or mechanical skimming [17]. (Fig.5)

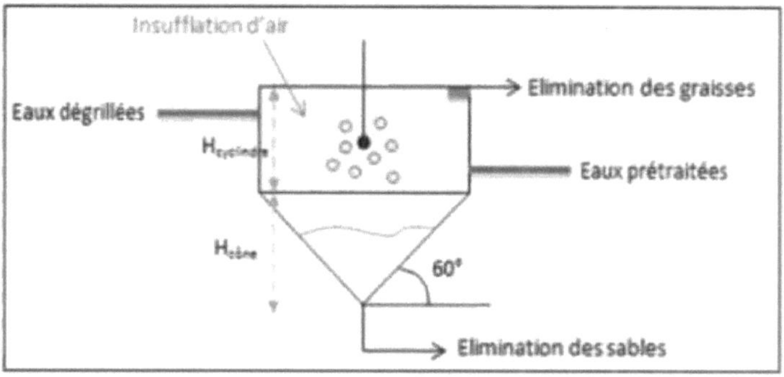

Figure 5 - Oil and grease removal.

b) Primary treatment (physico-chemical)

- ***Decanting:*** This is an essential process in primary treatment, the aim of which is to :
- retain a significant proportion of organic pollution.
- reduce the load on subsequent biological treatment.
- reduce the risk of clogging of biological treatment systems using fixed cultures (bacterial beds, biological discs, etc.).
- eliminate 30% to 35% of BOD and 60% to 90% of settleable matter (for domestic wastewater) [15].

There are 2 types of decanter:

- ***The conventional clarifier:*** eliminates at least 50 to 60% of settleable suspended solids in the raw water. It reduces the downstream dimensional characteristics of carbon pollution removal. This primary decantation process may be physical or physico-chemical, and the decanter may be rectangular, circular or lamellar in shape [17].

- ***The lamellar decanter:*** offers excellent purification performance, a better rise rate of 10 to 15 m/h at peak flow, a good residence time of 10 to 12 minutes, and takes up 6 times less space than the other two.

Its characteristics are better than those of rectangular and circular decanters, which are cumbersome, have a rise speed of 2 m/h and a residence time of 0.5 to 2 h [18].

- ***Actiflo process:*** The Actiflo process consists of organising precipitation on microsand grains, settling in a lamellar decanter and recycling the microsand after passing through a hydrocyclone. The system increases the contact surface, which facilitates the formation of flocs and accelerates their settling (Fig. 6). It operates at a rate of 60 m/h for primary treatment, 50 to 145 m/h for tertiary treatment, and 75 m/h for rainwater treatment [17].

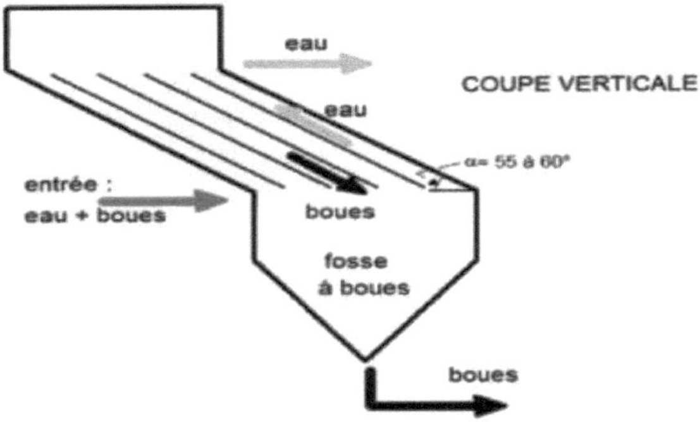

Figure 6 - Diagram of a lamellar clarifier.

- ***Procede Densadeg:***

The Densadeg process involves internal recirculation of a fraction of the sludge bed ranging from 100 to 200% (active sludge biological treatment), rich in nitrates. The sludge fraction is returned to the settling tank inlet, where it mixes with the coagulated water in the biological reactor to improve nitrogen and phosphorus treatment. The upward operating speed of the process is around 20 to 30 m/h in primary treatment and 17 to 25 m/h in tertiary treatment, which are lower speeds than for the Actiflo process, but can be as high as 40 to 120 m/h in mixed treatment for the Densadeg process [17]. (Fig. 7)

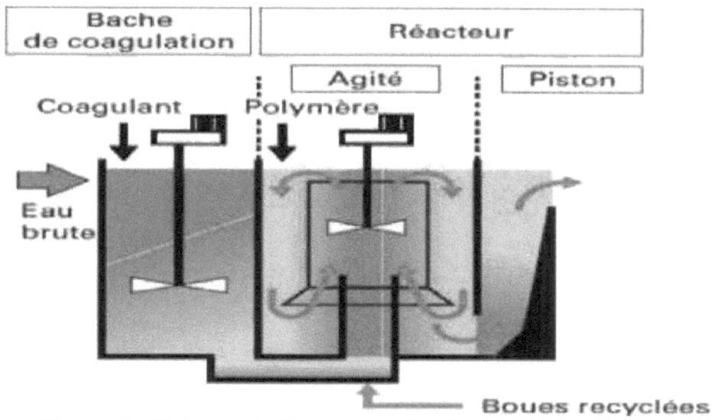

Figure 7 - Schematic diagram of the Densadeg process.

♦ ***Coagulation-flocculation***
Turbidity and colour in water are mainly caused by very small particles, known as colloids. To eliminate these particles, coagulation and flocculation processes are used: the main aim of coagulation is to destabilise the particles in suspension. The aim of flocculation is to promote contact between the destabilised particles by mixing them slowly.

c) Secondary treatment
Secondary treatment is most commonly carried out biologically. A physical-chemical process may replace it or, more often, be added to it to promote the flocculation and coagulation of the sludge or, for example, to allow the fixation of phosphates brought in by the use of fertilisers for agricultural activities.
The techniques most developed for urban wastewater treatment plants are intensive biological processes. The principle behind these processes is to localise and intensify the transformation and destruction of organic matter that can be observed in the natural environment.

1.5.2 Techniques with microbiological auxiliaries
The most widely used microbiologically-assisted techniques boil down to three processes: active sludge, bacterial beds and biological discs. Accelerated biological filtration (biofiltration) techniques are also used.

a) Active sludge WWTP
The principle behind active sludge is to intensify the self-purification processes found in natural environments.
The "active sludge" process involves mixing and agitating raw wastewater with bacteriologically highly active liquid active sludge. Aerobic degradation of the pollution takes place through intimate mixing of the purifying micro-organisms and the effluent to be treated. The "waste water" and purifying sludge phases are then separated (Fig.8).
An installation of this type involves the following steps:
- Preliminary treatments.
- The activation basin (or aeration basin).
- The secondary clarifier with recovery of some of the sludge.
- Disposal of treated water.
- Digesters for excess sludge from decanters.

besser deqraisseur

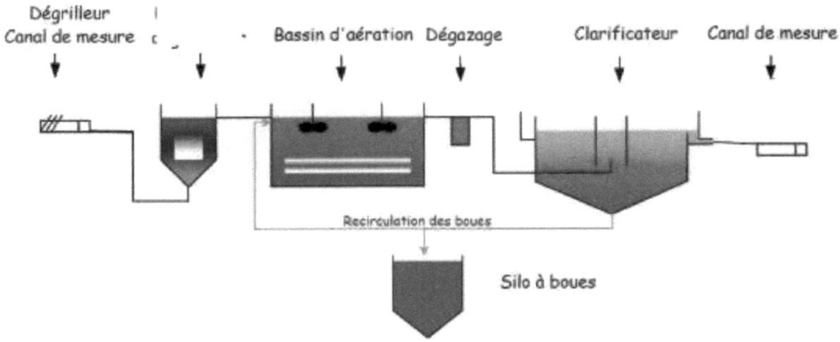

Figure 8 - Active sludge treatment process

b) WWTP with bacterial beds

The operating principle of a bacterial bed consists of allowing the wastewater, which has previously been decanted, to run over a mass of porous or cavernous material which serves as a support for the purifying micro-organisms. The bacterial bed consists of a cylindrical mass of pebbles or cobbles 5 to 10 cm in size. The layer height is between 1.5 and 2.1 m.

The bacterial bed must be equipped with a sprinkler system above the bed and a drainage and aeration system below the bed (Fig.9).

There are two types of bacterial beds:

- Single-stage bacterial bed.
- Bacterial bed with two filtration stages.

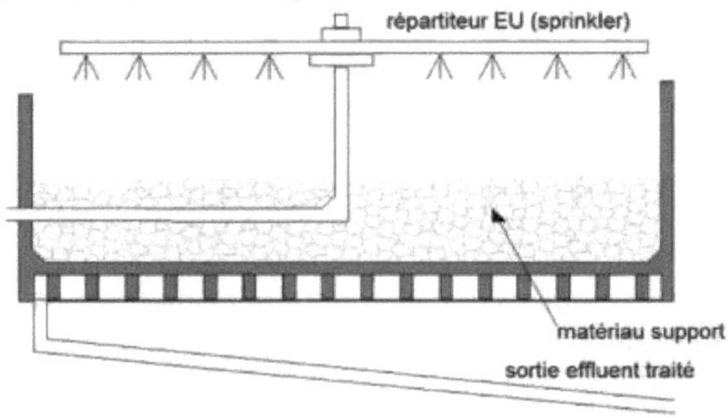

Figure 9 - Treatment process using a bacterial bed.

Aeration is provided either by natural draught or by forced ventilation. The aim is to provide the oxygen needed to keep the aerobic bacteria in good working order. The pollutants contained in the water and the oxygen in the air diffuse against the current through the biological film to the assimilating micro-organisms.

The biofilm contains aerobic bacteria on the surface and anaerobic bacteria near the bottom. The by-products and carbon dioxide produced by the purification process are released into the liquid and gaseous fluids.

c) Biological disc WWTP

Another technique using fixed cultures is the rotating biological discs. The micro-organisms develop and form a biological film on the surface of the discs. As the discs are semi-submerged, their rotation oxygenates the fixed biomass. For this type of installation, it is

important to ensure that :
- The mechanical reliability of the armature (soft-start drive, good attachment of the support to the shaft).
The sizing of the disc surface (this must be done with large safety margins).
The biological disc consists of a series of plastic discs with a corrugated surface approximately 3 metres in diameter, mounted on a horizontal axis. These discs are immersed to 40% in a basin containing the water to be treated. The discs are sufficiently spaced to allow the water to circulate freely.
The biological film covering the discs is alternately in contact with wastewater and air. The excess biomass in the discs is evacuated with the effluent and then decanted. This treatment process is recommended for towns of 300 to 2,000 population equivalent (Fig.10).

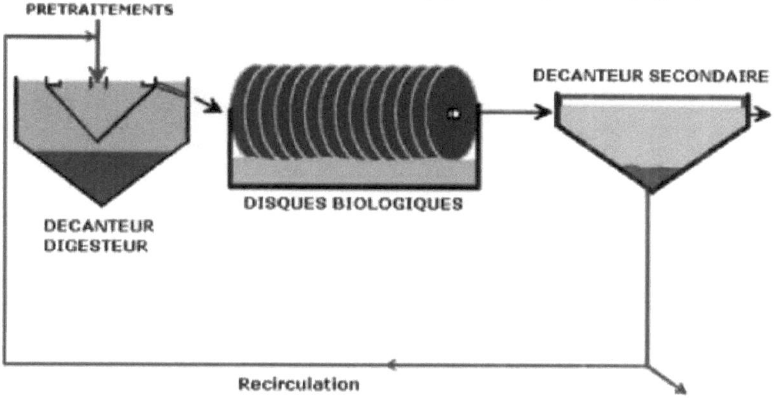

Figure 10 - Operation of a biological disc station.

Overall, the discs rotate slowly in the tank and, as they pass through the wastewater, the organic matter is absorbed by the bio-film attached to the rotating disc. The accumulation of organic matter on the discs increases their thickness and forms a layer of sludge. When the discs are exposed to the air, oxygen is absorbed, promoting the growth of this biomass.
When the latter is sufficiently thick (around 5 mm), a certain quantity is detached and settles at the bottom of the unit. The alternating phases of contact with the air and the effluent to be treated, combined with the rotation of the support, allow the system to be oxygenated and the bacterial culture to develop.
During the submerged phase, the biomass absorbs the organic matter, which it degrades by aerobic fermentation using atmospheric oxygen from the emerged phase. The materials used are increasingly lightweight (generally expanded polystyrene) and the actual surface area developed is increasingly large (flat or alveolar disc).
The advantages of the biological disk purification process are :
- The ease of operation.
- A high percentage of BOD elimination.
- Good decantability of the sludge produced.

I.5.3 Tertiary treatment

Some treated water discharges are subject to specific regulations concerning the elimination of nitrogen, phosphorus or pathogenic germs, which require the implementation of tertiary treatments [20]. This covers all the physical and chemical operations that complete the primary and secondary treatments.

a) Elimination of nitrogen :

Wastewater treatment plants eliminate only around 20% of the nitrogen present in wastewater through nitrification-denitrification treatments.

To meet discharge standards in sensitive areas, additional physical and physico-chemical processes can be used to eliminate nitrogen by electrodialysis, ion exchange resins and ammonia stripping, but these treatments are not used to treat urban wastewater, for reasons of efficiency and cost [20].

Nitrogen is generally eliminated by a biological process in two important stages.

♦ **Nitrification**: this is a process that takes place under the action of certain specific micro-organisms and leads to the transformation of ammonia (or ammonium) into nitrate in two (2) stages:

- Nitrosation by aerobic nitrous bacteria (nitrosomonas)
- Nitration by aerobic nitrifying bacteria (nitrobacter).

Nitrification is one of the stages in the treatment of wastewater, which aims to convert ammonium (NH4) into nitrate (NO). This transformation is carried out by bacteria in an aerobic environment.

♦ **Denitrification:** this **is** an anaerobic process by which nitrates are reduced to nitrogen and nitrogen oxides. The micro-organisms use the nitrates as a source of oxidant instead of oxygen and in the presence of a source of an organic carbon that must be brought into the environment [21].

b) Phosphorus elimination :

Phosphorus removal, or dephosphorisation, can be achieved by physical-chemical or biological means. In the case of physico-chemical treatment, radioactive or reactive substances such as iron or aluminium salts can be used to precipitate insoluble phosphates and eliminate them by decantation.

These techniques, which are currently the most widely used, eliminate between 80% and 90% of phosphorus, but produce a lot of sludge [21].

♦ **Odour elimination and treatment**

Wastewater, loaded with particulate and dissolved organic matter, can directly or indirectly induce the formation of unpleasant odours through its purification by-products (grease, sludge) as a result of a fermentation process.

Odours from WWTPs are due to gases, aerosols or vapours emitted by certain products contained in the wastewater or in the compounds formed during the various treatment phases.

The main sources of odours are :

- Pre-treatment.
- Sludge and its treatment.

To avoid these nuisances, sensitive structures will be covered and fitted with a ventilation system and a biological odour treatment unit. There are generally two types of biological odour treatment: bio-filters and bio-scrubbers.

In the first, the biomass is supported by a specific floor and the air passes through the mass (often peat). The latter create a second filter using a suspension. The biomass is free, and purification takes place in a reactor [22].

♦ **Disinfection:**

A reduction in the germ content, sometimes required for discharges into specific areas (bathing areas, shellfish farming areas) or for reuse, will be achieved by chemical disinfection treatments using :

- ***Chlorine*** (Cl): is a powerful oxidant that reacts both with reduced and organic molecules, and with micro-organisms. The compounds used in wastewater treatment are: chlorine gas (Cl_2), sodium hypochlorite (NaCIO), commonly known as bleach, calcium hypochlorite ($CaCIO_2$, chlorine lime (CaCI,OCI) and sodium chlorite ($NaClO_2$).
- ***Ozone (o3)*:** is a powerful oxidant. Disinfection using FO is used in

It is mainly used in the United States, South Africa and the Middle East. It eliminates bacteria, viruses and protozoa. It is the only process that is truly effective against viruses. Toxicity tests carried out on fish, crustaceans and algae have not revealed any toxicity.

There are also physical treatments, such as :

- ***Ultraviolet rays***: using mercury lamps positioned parallel or perpendicular to the water flow. Their radiation attacks the micro-organisms directly. This treatment is very simple to implement because there is no storage or handling of chemical substances and the chemical characteristics of the effluent are not altered.
- ***Filtration***: is a physical process that allows micro-organisms to be retained using a filter. Whether it is carried out on sand or on a membrane, this technique requires secondary purification beforehand to guarantee fairly thorough elimination of suspended matter. The removal of viruses, bacteria and protozoa depends on the porous medium, the percolation rate, the thickness of the filter bed and the level of oxidation of the filtered water [21].

I.6 Sludge treatment

Sludge is the main residue from wastewater treatment plants. Sludge treatment represents 30% of the investment in the construction of a wastewater treatment plant. The purpose of sludge treatment is to :

- ♦ Reduce the organic fraction in order to reduce its fermentability and the risk of contamination (stabilisation).
- ♦ Reduce their total volume in order to reduce disposal costs (dehydration).
- ♦ Final disposal of sludge by : [1]

- Agricultural development.
- Incineration.
- Landfill.

I.7 Natural lagooning

Lagooning is an extensive type of wastewater treatment process, in which the treated water is left to soak for a short time, thereby providing space and sunlight. It is equipped with several shallow tanks (at least three) where bacteria and other living organisms proliferate to the detriment of the organic matter and mineral salts contained in the water. The number of pathogenic agents (bacteria, viruses, parasites) is considerably reduced, in particular because of the long retention period in the reservoirs.

This is a biological purification process that eliminates biodegradable organic matter and produces mineral salts, which can lead to a proliferation of algae (phytoplankton) that grows under the combined effect of nitrogen and phosphorus derivatives in the water and photosynthesis due to solar radiation (eutrophication).

However, this phenomenon, which is very harmful to natural waters, proves to be beneficial in the lagooning process. In fact, the destruction of organic matter takes place thanks to an extremely wide-ranging biological association:

- At the bottom of the first lagoon, a category of micro-organisms degrade the organic matter using classic anaerobic fermentation processes.
- In the upper part of the head pond and in the other ponds, aerobiosis is widespread and bacteria mineralise the soluble organic matter in suspension.

The metabolism of these aerobic bacteria requires a large supply of oxygen from the photosynthesis of algae, which grow vigorously in a nutrient-rich environment (Fig.11).

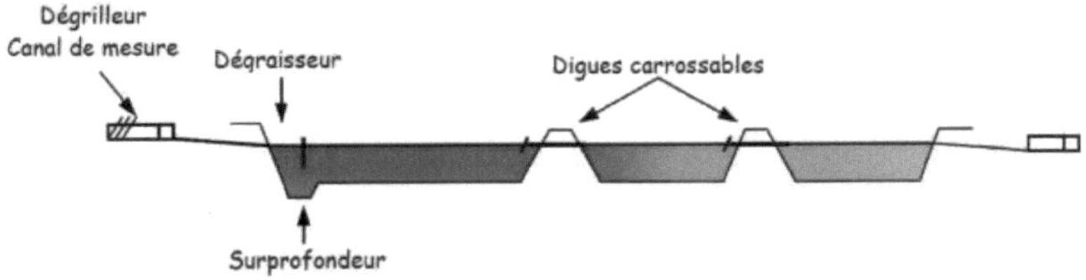

Figure 11 - Schematic diagram of a lagooning plant

1.7.1 Operation of a natural lagooning plant

Basin no. 1: Anaerobie

Bacteria are unicellular, prokaryotic micro-organisms (a single cell without a nucleus). They generally reproduce by simple cell division and are able to resist unfavourable conditions in the form of spores. There are two types of bacteria:

- *Anaerobic bacteria*: can develop in the absence of oxygen.
- *Aerobic bacteria*: they need oxygen to live.

In lagoons, the former are found at the bottom of the pools and in the mud, while the latter dominate in the open water. They mainly use the oxygen generated by the activity of microscopic algae suspended in the water for their respiration. The micro-organisms are eliminated by a process known as mineralisation (Fig.12).

Organic materials Mineral materials

This mineralisation of organic matter by different processes produces water, minerals (NHA NO_2 . NO, SO, PO) and gases (CO2, H_2 S, CH4, NH....h) which are gradually released into the second basin [23].

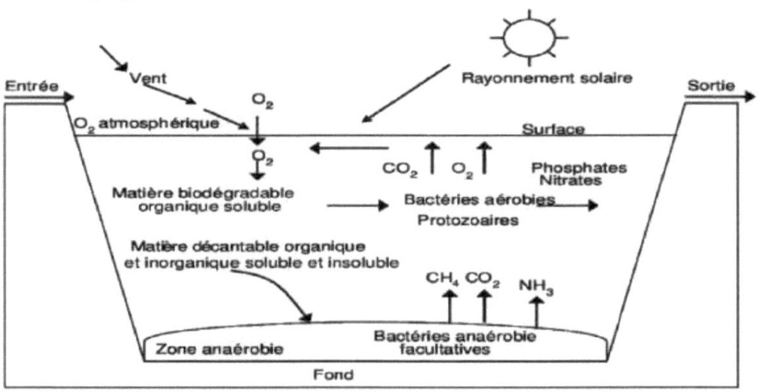

Figure 12 - Operation of an aerobic lagoon

Basin no. 2: Optional

Once the bacteria have broken down the organic matter, the plants come in to fix the products of mineralisation. The water then flows into this second basin, which is half the size of the first and less deep. This shallower depth allows good photosynthesis, a more effective role for ultraviolet rays, but above all, it encourages aerobic phenomena.

The nutrients present (mineral salts, laundry by-products and, to a lesser extent, mineral fertilisers from agriculture) and the CO_2 (waste from the respiration of certain bacteria) will be assimilated by the plants to enable them to grow. These autotrophic organisms transform the various mineral salts and CO2 directly, using solar energy, into organic tissue (sugars) for the plant and oxygen for release into the environment: this is the phenomenon of photosynthesis.

$$\textbf{Mineral salts} + CO_2 + H_2O\ \textbf{glucose}\ \text{(algal growth)} + H_2O + O_2$$

Basin no. 3: Maturation

They also play an important role as consumers of microalgae, and therefore as regulators of these phytoplankton populations.

► **Protozoa**: these unicellular organisms are the main predators of bacteria. They are present all year round without showing any major numerical evolution *(paramecies, vorticella, epidiscal)* [23].

► **Metazoans** these multi-cellular organisms, of greater complexity, are present in the last basins of the lagoon plant under three dominant groups

- ***The rotifers:*** These are essentially microphages that consume bacteria, microalgae and organic matter to clarify water effectively.
- ***Copepods:*** These are small crustaceans (0.5 to 4 mm) at the surface of the water and are very efficient predators: they eat phytoplankton, young insect larvae and cladocerans.
- ***Cladocerans:*** These are small herbivorous and detritivorous crustaceans measuring 0.2 to 3 mm. They play an important role in the lagoon plant and partially in the final basins to reduce the suspended matter content [23].

1.7.2 Constraints and malfunctions in the lagoon system

The main causes of failure in a lagoon-based wastewater treatment system are a lack of oxygen and its poor distribution in the water column. The lack of oxygen or the development of the anoxic zone can be caused by : [24]

- Decrease in algal growth due to lack of light penetration. This phenomenon can occur in autumn or winter when sunlight is more limited.
- Formation of a thermocline, particularly during the summer period, limiting gas exchange on either side of the stratification.
- Decrease in the concentration of dissolved oxygen, which occurs during the summer period, as the temperature (and salinity) increases and the concentration of dissolved oxygen decreases.
- Decrease in the dissolution of air oxygen in water when winds are limited; this situation occurs mainly during the summer.
- An increase in the load of organic matter promoting the growth of heterotrophic bacteria that consume oxygen; if this load becomes too high, this can lead to deoxygenation of the water.

In order to limit the problems caused by a lack of oxygen, mixing the water may be the appropriate solution. Among other things, it makes it possible to :

- *Eliminating foul odours:* Efficient mineralisation of biodegradable organic matter suspended in the water column leads to a reduction in the volume of biodegradable organic matter deposits and therefore a reduction in the release of fermentation products (methane, carbon dioxide).

- *Degrading pollutants:* The organic load removed from the environment decreases when the heterotrophic bacteria do not have sufficient oxygen to degrade the organic matter. This results in increased deposition of organic matter on the bottom and BOD5 discharges that are higher than they should be.

Chapter II:

PHYTODEPURATION

Introduction

This is a natural lagoon system, with a more pronounced role for the vegetation used to purify the wastewater. The choice of plants can vary greatly depending on hydro-climatic and edaphic factors. The advantages of this approach are ecological, economic and aesthetic. There are two types of phytodepuration, classified according to the type of vegetation [23].

- ***Microphyte lagooning:*** The only plants are microscopic algae averaging 1/100 mm in size, but which play the same role as macrophytes in fixing nutrients.
- ***Macrophyte lagooning (or phytoëpuration):*** characterised by the presence of plants immersed in the water. It consists of shallower, shallower filter basins planted with aquatic plant species, capable of extracting pollutants from wastewater with a relatively low pollution load [23].

II.1 Historical background

Phytoremediation is not a new concept; 300 years ago, plants were already being used to treat water and eliminate, contain or render less toxic environmental contaminants.

In the 16th century, Andrea cessalpino, a botanist from Florence, discovered a plant growing on rocks that were naturally rich in metals (particularly nickel). From 1814 to 1948, numerous studies were carried out by scientists on this plant, called *Alysum bertolonii,* and it was discovered that it accumulated in its body a high level of metals from the soil in which it lived, a level higher than that found in the soil itself.

Phytoremediation is a low-cost alternative or complement to intensive and expensive treatment technologies. Moreover, unlike technologies that simply move pollutants, vegetation offers a means of totally eliminating waste from soil and water, since, in the rhizosphere, organic pollutants are totally destroyed (mineralised) [25].

Later, other plants with the same properties were discovered. And it was in 1970 that the idea of using these plants with their particular properties emerged. Cunningham et al (1995) defined 'phytoremediation' as a technique involving the use of green plants and their associated micro-biotopes, soil amendments and agricultural techniques to eliminate, contain or render less toxic environmental contaminants.

They add that there is also 'plant-assisted bioremediation', in which plant roots and their rhizosphere microorganisms are used to remediate soil and water contaminated with organic compounds.

Ernst in 1988 confirmed that the choice for the elimination of toxic metals from the soil (heavy metals and radionuclides), calls for the use of selected plants that accumulate these metals. This technique comprises three aspects: phytostabilisation, phytoextraction and rhizofiltration [26].

II.2 Definitions and concepts

This is a wastewater treatment system that uses plants (usually macrophyte plants), substrates and micro-organisms in an artificial wetland. Phyto-purification systems can consist of one or more plant filters. Unlike the macrophyte lagoon, the plant filter uses a filter substrate (sand, gravel, gravel). There are two types of filter:

a) Vertical flow filters

The water is distributed over the surface of the filter and percolates through the gravel pack. There is therefore no stagnation of water or photosynthesis, unlike in a lagoon. Microbial activity is heavily involved.

The plants preferred for this type of filter are generally aquatic plants with strong root systems (irises, cattails), as their rhizosphere promotes air circulation during dry periods and helps to maintain the hydraulic conductivity of the filter bed over time (unclogging). This intensifies bacterial activity.

The organic matter retained on the surface becomes moist and mineralised. The bacterial flora used in this filter is therefore of the heterotrophic aerobic type, which settles on the medium.

b) Horizontal flow filters

The horizontal flow filter has a flow similar to that of a lagoon, since a water level is maintained in the basin, and excess water leaves the basin via an overflow system.

However, the tank also contains a filter medium (sand and gravel), and the water level is lower than the sand layer. No photosynthesis takes place in this filter.

The bacterial flora settles on the granular medium, and anoxia phenomena can favour partial denitrification if this filter is used after a vertical flow filter. In addition, phosphorus, pharmaceutical molecules and hormones may be reduced (Fig. 13).

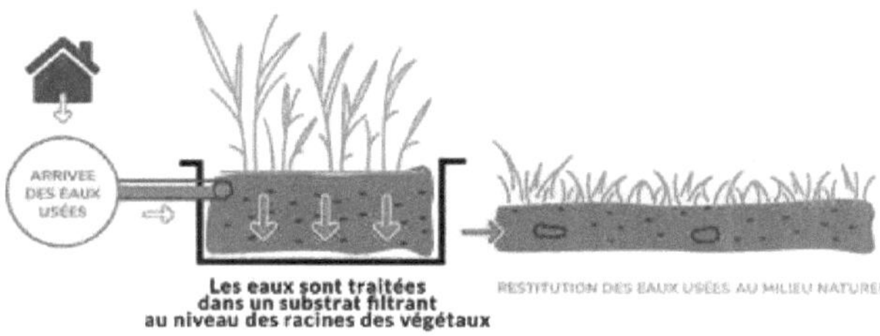

The water is treated AND RETURNED TO THE NATURAL ENVIRONMENT **in a filter substrate.**
in the roots of plants

Figure 13 - Cross-section of phytodepuration process

11.3 Reed filters (FPR)

Macrophyte plant filter technology for wastewater treatment in small communities is a recent development. The first research by biologist Kate SEIDEL (1946) was based on rigorous observation of the life of marsh and wetland plants. They revealed that :

J The installation of helophytic plants (reeds, cattails, rushes, etc.) in an entirely mineral substrate irrigated by waste water, leads to a purification process.

J The rhizomes and roots of these plants provide a highly favourable environment for the activity of micro-organisms that degrade organic matter. The leaves, stems and rhizomes of macrophytes provide these micro-organisms with oxygen and the various supplements they need to thrive.

J The combination of different types of flow (vertical or horizontal) enables degradation and partial mineralisation of the organic matter, as well as denitrification, in aerobic and anaerobic

phases.
It is a reliable technique that is simple to operate, greatly facilitates sludge management and is well accepted by local residents because of its perceived "natural" image, reinforced by its ability to blend into the rural landscape **[27].**

A wetland is a purification station in itself, through its natural action of degrading and eliminating organic matter. This self-purification process is largely due to living organisms (bacteria, algae) which mineralise the organic matter, which is then assimilated by the higher vegetation (macrophytes).

Some of them, and reeds in particular, oxygenate the environment and thus encourage the development of aerobic micro-organisms (in the presence of air and oxygen in particular).

Techniques such as natural lagooning can be integrated into very simple systems such as "planted infiltration ditches", which allow rainwater run-off to be returned to the water table while being treated at the same time. The use of "rustic processes" such as reed filters seems to be a good alternative to conventional treatment plants for these small communities (Fig. 14).

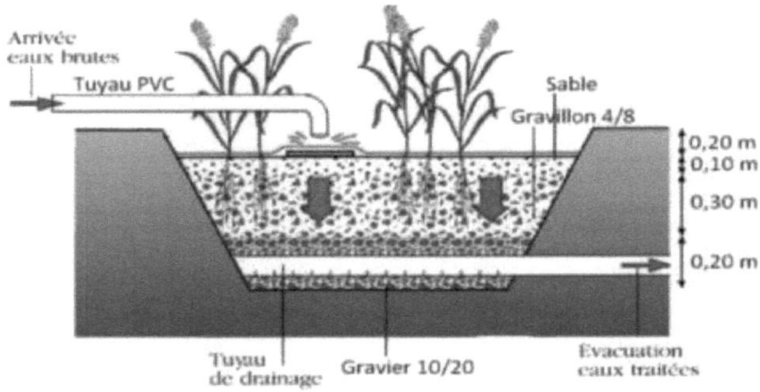

Figure 14 - Diagram of a vertical flow reed filter

II.4 Operation of wastewater treatment plants

This is a biological process with fixed cultures on fine supports, based on the passage of wastewater through filter beds colonised by bacteria, which carry out the purification processes. The wastewater is kept below the surface of the plant beds and circulates in two types of flow (vertical or horizontal), flowing by gravity down a natural slope.

The aim is to combat the polluting effects of water discharged daily from our rural towns and industrial plants. This type of purification can be applied to several types of water: raw wastewater, decanted wastewater, biologically treated water and rainwater.

It is an ecological process that is rapidly expanding in developed countries, due to its reliability, simplicity and rusticity (requiring no energy). It is also well accepted by the local population because of its natural appearance, perfectly integrated into the rural landscape.

Nowadays, experiments are multiplying, taking into account not only the objectives of treating polluted water, but also the variations in physical and geographical conditions in each area.

The introduction in Algeria of this type of extensive purification technique could provide commendable services to local communities in the face of the problems of environmental degradation in general and the recycling of water in certain agricultural and forestry activities.

11.5Role of the various process components

- Role of the filler material

- By virtue of its particle size, the filling material has an obvious role in filtering the suspended matter present in the wastewater, hence the name "filter". Its effectiveness in this role depends to a large extent on the texture of the material, which is approached by its granulometry and which will have a particular influence on the hydrodynamic characteristics (hydraulic conductivity in a saturated or unsaturated environment).
- The composition of the filling material also influences the treatment through its ability to absorb phosphorus and heavy metals. This depends essentially on its iron, aluminium and calcium content, and the residence time of the water in the bed, which varies according to the porosity of the material in place, in relation to the kinetics compatible with the treatment objectives set.

- The role of plants

- Aquatic plants have developed the ability to transfer the oxygen formed by chlorophyll synthesis to their underground parts (rhizomes, roots and rootlets). Some of this oxygen is excreted into the surrounding environment to help oxidise the mineral salts needed for plant nutrition.
- The roots, like the mineral substrate, support microbial development in their underground parts. The microbial populations present in the support material and in the rhizosphere (rhizome and root zone) are greater than in non-plant filters. They assimilate certain substances, such as nitrogen and phosphorus, for their own metabolism and/or to store them. Nitrogen elimination is linked to the plant biomass produced.
- Finally, it is thought that certain plants secrete antibiotics in their roots, thereby helping to eliminate pathogenic micro-organisms [28].

- Role of micro-organisms

The main role of micro-organisms is to degrade organic matter. They carry out the various oxidation and reduction processes. They mineralise nitrogen and phosphorus compounds, making them available to plants. They also carry out nitrification/denitrification reactions. The anoxic bacteria present under the aggregates eliminate the nitrates in the form of nitrogen gas (Fig.15).

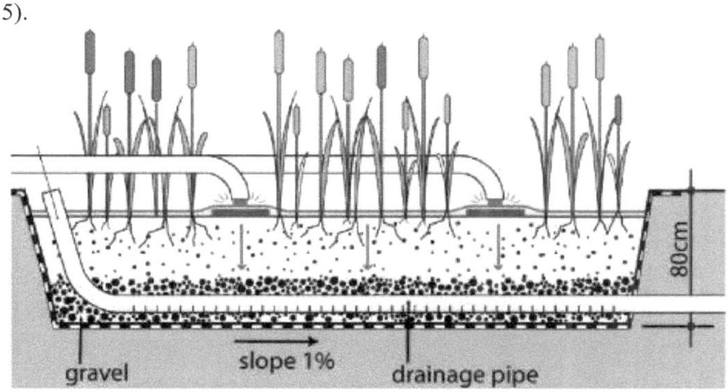

Figure 15- Components of a macrophyte plant bed

There are two types of filter: vertical and horizontal. These two filters complement each other quite well and are generally used in series, forming a more efficient system (Fig.16).

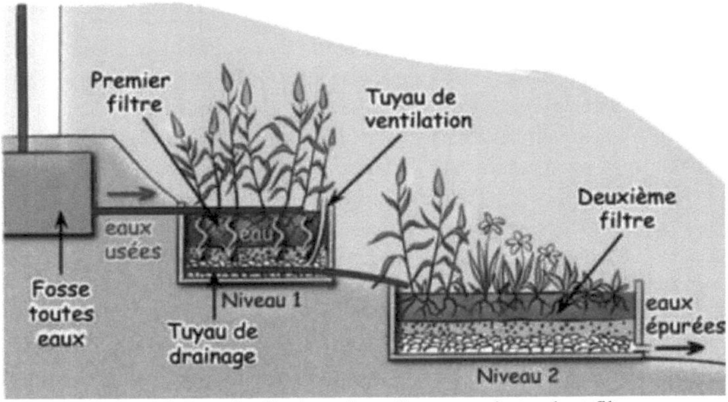

Figure 16- Longitudinal section of a macrophyte plant filter.

II.6 Choice of plant species

Plants used in phytoremediation must meet certain conditions: [29]

- Firstly, it must be able to tolerate high concentrations of the element to be removed from the water in the cells of its roots and aerial tract. Hyper-tolerance is the key property that makes this hyper-accumulation possible;
- Be adapted to the climatic and pedological conditions of the site;
- Have a well-developed root system to effectively stabilise the soil;
- The plant must be capable of translocating the element in question, at a high rate, from its roots to its aerial tract, and then storing it;
- The plant must have a well-developed root and leaf system.

The choice of species depends on the balance between what is expected of the plant and its ability to grow on the site.

The growth of roots and rhizomes maintains or regulates the initial hydraulic conductivity. The low granulometry of the substrate (sand or gravel) and the high input of organic matter are conducive to clogging. Root growth limits these risks by forming tubular pores along the growing roots [30].

Above all, root development increases the fixation surface for the development of micro-organisms. They also contribute to the biological degradation of the organic part of TSS and release low molecular weight organic compounds into the environment, which can be polymerised and easily degraded [31].

In addition to this increase in active surface area, there is undoubtedly an as yet poorly documented factor that stimulates the activity, and even the diversity and density, of the micro-organisms involved in various ways in the purification processes. This concept is well known in agronomy and can be summed up in the trivial form 30

This is sometimes referred to as the *"rhizosphere effect"*.

Micro-organisms are very abundant in the soil and particularly active in the rhizosphere in connection with the exudation of organic carbon for the roots. It was the stimulation of their activity that gave rise to the concept of the rhizosphere. They can have negative, positive or neutral effects on root growth, depending on the species and environmental conditions [32].

There are several types of vegetation linked to the presence of water. These include unistrate and pluristrate aquatic herbaceous vegetation, semi-aquatic herbaceous vegetation and the various types of shrubby and arborescent vegetation associated with hydromorphic soils (riparian forests, forest galleries) [33].

a) Hydrophytes: These are plants that colonise the interior of bodies of water or watercourses. They can be :

- Outcrops, which are initially completely submerged at the start of vegetation, then rise and reach the water's surface;
- Floating, freely on the surface or between two waters;
- Completely submerged, forming veritable meadows at the bottom of the water.

The most common hydrophytes encountered include [33] :

- *Lemna, polyrrhiza minor*	Water lentil with many roots
- *Salvinia natans*	Enlarged leaves and branching roots
- *Polygonum anphibium*	Amphibious knotweed
- *Potamogeton natans*	Heart-shaped oval-leaved pondweed
- *Ranonculus divaricatus*	Twig-leaved Crowfoot
- *Elodea sp.*	Elongated elliptical leaves, whole

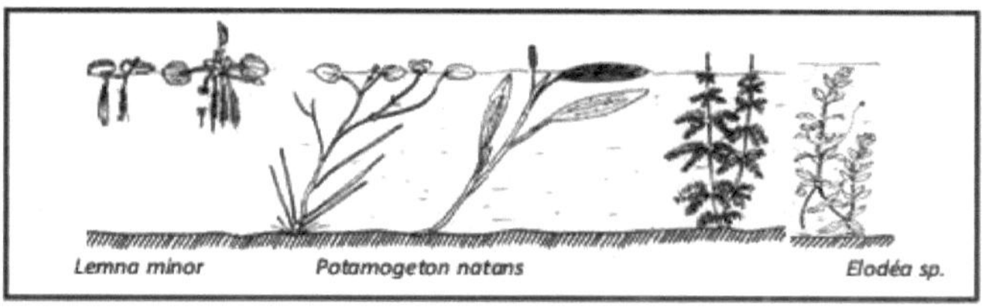

b) Amphiphytes: These are amphibious plants, capable of settling and developing below and above water surfaces. The main species are :

- *Ludwigia grandiflora* and *L.* peploidesthe primrose willows
- *Myriophyllum brasil* Brazilian water milfoil

c) Helophytes: These are plants that colonise the edges of bodies of water and watercourses, whose root system is located in the water-logged substrate but whose vegetative apparatus is located outside the water. These are reedbeds, cariga and other groups of large herbaceous shore plants. Some water helophytes of ponds [34] :

Carex sp.	Laiches
Epilobium sp.	Fireweed
Iris pseudaconus	Water iris
- *Juncus sp.*	Bulrushes
- *Mentha aquatica*	Water mint
- Thalaris arundinacea	False reed, baldingere Rubanier.
- Sparganum sp	

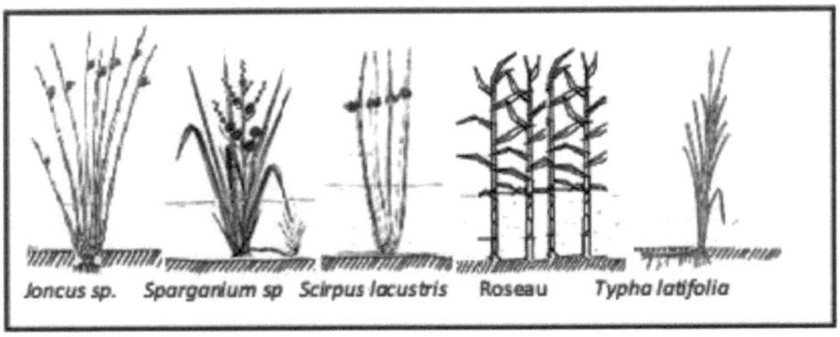

With a view to long-term management of water resources, vegetation can be a control tool that should not be overlooked.

For proof of this, we need look no further than the research work currently underway to develop a plant-based bio-indication method. It has been established that they reflect the conditions of the environment or its evolution.

What's more, their development is not simply a consequence of this, but can have a knock-on effect, particularly on the physico-chemical quality of the water and its suitability for certain uses.

However, in order to include the plant compartment, it will be necessary to complete the following steps

the inadequacy of current knowledge on certain specific points.

Firstly, the lack of precise quantitative information means that campaigns will have to be set up to measure the plant potential present in the watercourses.

Chapter III

EXPERIMENTAL PILOTS

Introduction

In Algeria, few attempts have been made on the investment front, preferring to mature research on experimental pilots. Our laboratory has carried out two such studies:

- A scale model set up at the experimental farm of the University of Mascara;
- A full-scale pilot project at Brezina (El Bayadh).

In both cases, the TMEs are stored in a finishing tank before being recycled for crop irrigation. With its significant and encouraging results, the phytodepuration process offers a perfectly justifiable opportunity to contribute to the challenge of saving water and managing it efficiently, particularly in small communities.

III.1: In-situ experimental pilot

The experimental pilot plant for phytoremediation was installed at the Mascara university. Polyethylene basins and plants (reeds and cattails) were prepared, and wastewater samples were tested for treatment. In view of the covid pandemic, the residence time was one week. Finally, a comparative analysis was carried out between the two sites (lagoon plant and experimental pilot plant).

111.1.1 Description of the driver

a) Operating diagram

The principle adopted for setting up the experimental pilot plant is to reproduce the wastewater treatment stages of the El Keurt village lagooning plant, introducing the phytodepuration system, as shown in the diagram below (Fig. 17).

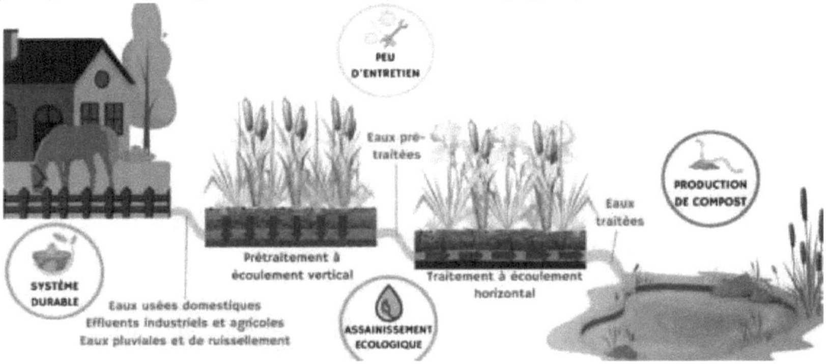

Figure 17- Schematic representation of a phytodepuration model

Our reduced pilot plant consists of a series of stepped basins planted with vegetation, allowing wastewater to flow by gravity from top to bottom. At the inlet, the raw wastewater (taken from the El Keurt pilot plant) is discharged. At the outlet, the treated water is collected for laboratory analysis.

b) The stages involved in setting up the experimental pilot

1°- Installation of four basins planted individually with macrophyte species (phragmites, cattails, rushes) + mixed microphyte species (iris, duckweed, cattails);

2°- Passage of raw used water taken from the El Keurt lagoon;

3°- Analysis of raw wastewater (pilot plant inlet) and wastewater treated by vegetation (pilot plant outlet) ;

4°- Comparison of results and interpretation in relation to regulatory standards for domestic wastewater discharges;

5°- Comparative analysis in relation to the reference lagooning system.

The purification principle of this process consists of allowing wastewater to pass through the root system, benefiting from natural oxygenation via the aerial stems. The result is microbial digestion and natural purification. At the end of the chain, the treated water is reused in agriculture to the required standards (Fig. 18).

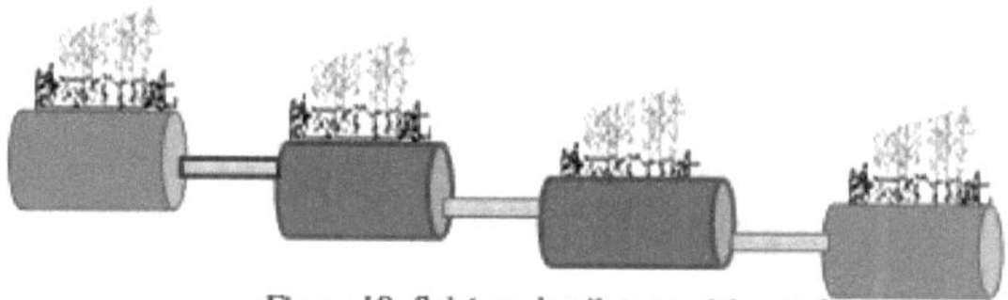

Figure 18- Schéma du pilote expérimental

Figure 18- Diagram of the experimental pilot

The final phase of the project consists of monitoring the parameters relating to :

- hydraulic measurements: inflow and outflow, water level in basins, retention time ;
- water quality on entering and leaving the pilot plant (temperature, pH, conductivity, suspended solids, BOD5, COD, ammonium, phosphorus, potassium, heavy metals, pathogens, etc.).
- growth factors for macrophyte plants: density, root development and activity, behaviour towards pollutants.

In each basin, it is assumed that the wastewater undergoes partial treatment, passing through the plant life for a residence time of 24 hours. By the time you reach the fourth tank, the water quality is clearer, leading to a physical reduction in pollution.

111.1.2 Analysis and results

***Temperature*:** The instrument used to measure temperature was an OX 192 oximeter fitted with an electrode. The results obtained are summarised in Table 4.

Table 4 - Temperature results (April 2022).

Samples taken	Raw water	Drained water
Natural lagooning	16.8	16.7
Macrophyte lagooning	16.8	13.2

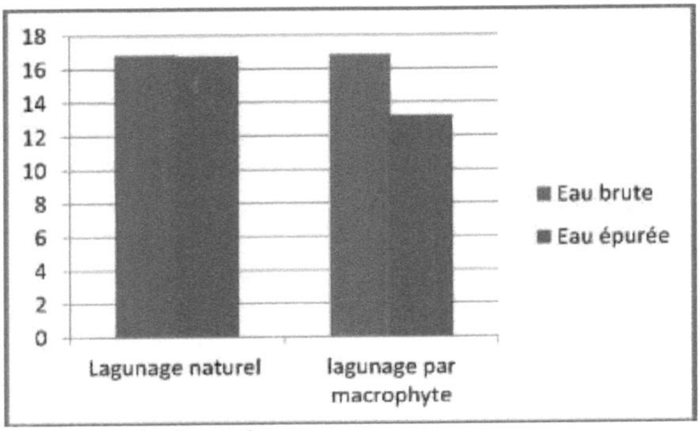

Figure 20 - Temperature analysis results.

We observe that the inlet of this water station is a little warmer than the outlet of the same station (El Keurt), and for our system the temperature is a little less warm (15.4-15.2 C°), due to the involvement of vegetation in the pilot station.

***pH*:** The pH of the water was measured using a portable pH meter, and the results obtained are summarised in Table 5.

Table 5- pH results (Avri 2022).

Samples taken	Raw water	Drained water
Natural lagooning	7.86	8.04
Macrophyte lagooning	7.86	7.92

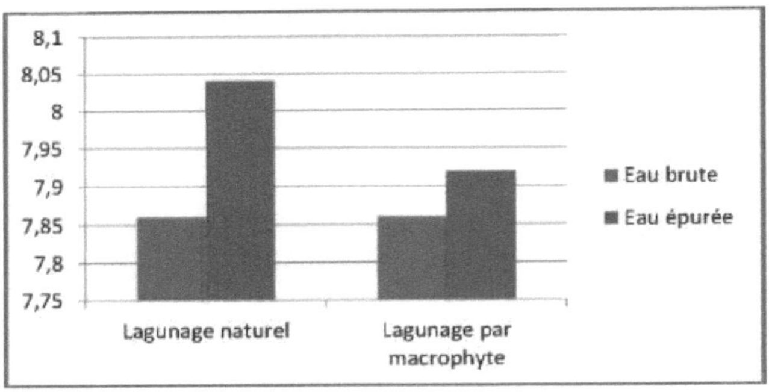

Figure 21- pH analysis results.

The pH values obtained from the El Keurt plant and the macrophyte process are close, between 7.77 and 8.04, which is an acceptable level compared with the standards.

Conductivity: This parameter reflects the overall mineralisation of the water, and was analysed using a portable conductivity meter. The results are summarised in table 6.

Table 6- Conductivity results (April 2022)

Samples taken	Raw water	Drained water
Natural lagooning	4380	3610
Macrophyte lagooning	4380	2877

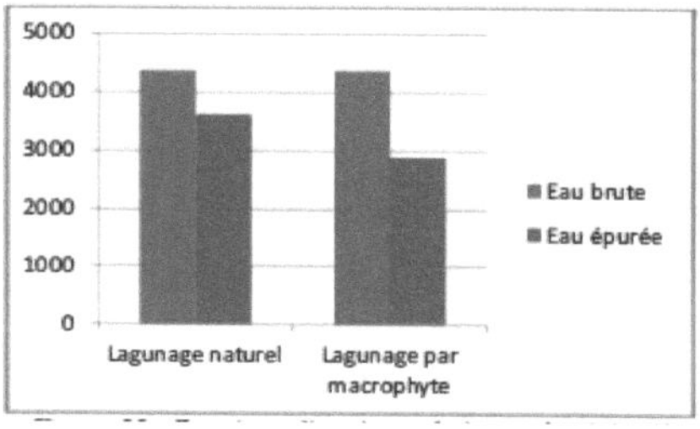

Figure 22 - Results of electrical conductivity analyses.

In both cases, the mineralisation process has evolved, and after treatment the wastewater is demineralised due to the presence of macro-biological vegetation in the pilot plant.

TSS: This is the quantity of organic or mineral matter suspended in the water. The results obtained are summarised in table 7.

Table 7- Suspended matter results (April 2022)

Samples taken	Raw water	Drained water
Natural lagooning	45	28
Macrophyte lagooning	45	22

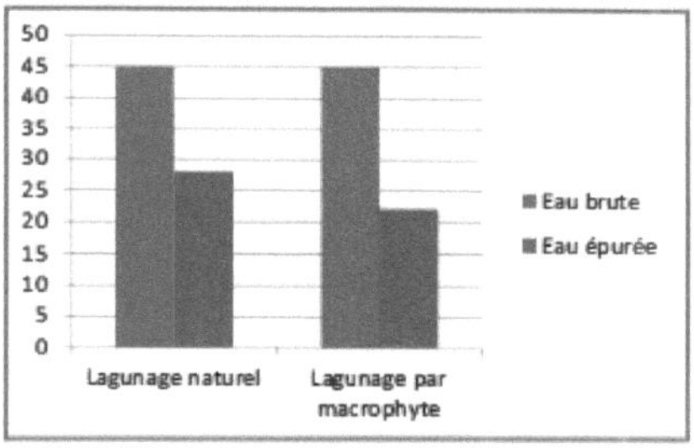

Figure 23 - TSS analysis results

TSS levels are very high at the inlet, at around 45 and 61 mg/l. At the outlet, they are close to standards, at around 28 and 54 mg/l, due to the settling of TSS in the plant's large basins, whereas in our pilot plant, TSS is eliminated by the reeds.

Dissolved oxygen : The presence and quantity of gaseous oxygen O2 dissolved in water is related to the temperature and volume of water. The results obtained are summarised in Table 8.

Table 8 - Dissolved oxygen results (April 2022)

Samples taken (mg/l)	Raw water	Drained water
Natural lagooning	0.6	1.11
Macrophyte lagooning	0.6	3.8

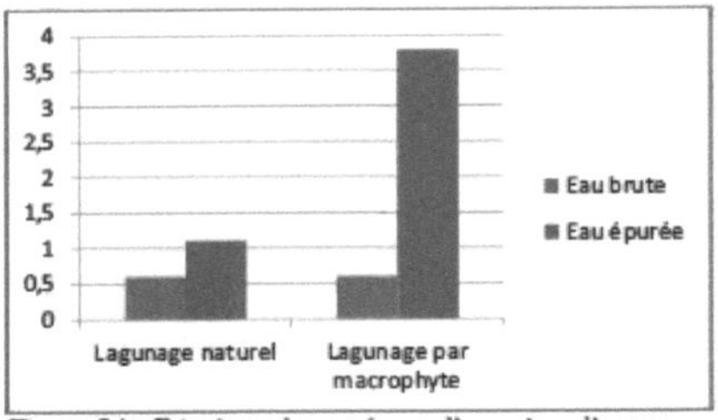

Figure 24 - Résultats des analyses d'oxygène dissous.

Figure 24 - Results of dissolved oxygen analyses.

According to the results obtained, the quantity of oxygen is low due to the presence of micropollutants in the water, but there is a small increase at ГO? of the pilot station due to a photosynthetic respiratory metabolism resulting in the release of oxygen into the aquatic environment by the macrophyte reeds.

Biological oxygen demand (BODs): Over a period of 5 days, the amount of oxygen consumed by the micro-organisms under the defined conditions was measured. The results obtained are summarised in Table 9.

Table 9- BOD results

Samples taken (mg/l)	Raw water	Drained water
Natural lagooning	230	75
Macrophyte lagooning	230	115

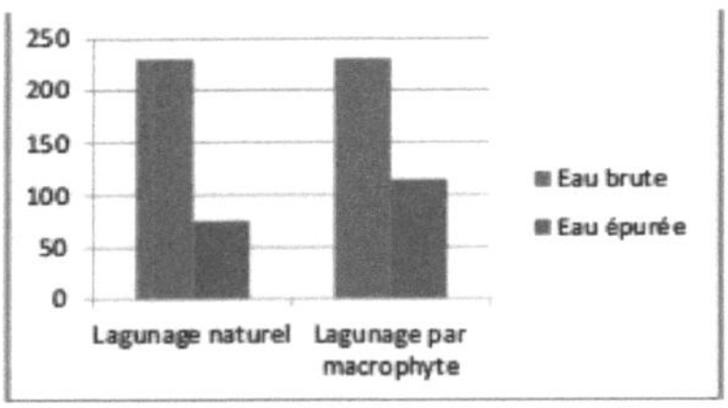

Figure 25 - Résultats des analyses de DBO_5.

A reduction in the values was achieved by eliminating the pollutant load using planted vegetation in the experimental pilot.

Chemical oxygen demand (COD): This chemical reaction, which indicates the amount of oxygen consumed by the oxidisable materials, takes place in a concentrated sulphuric medium at a high temperature. The results obtained are summarised in Table 10.

Table 10- COD results.

Samples taken (mg/l)	Raw water	Drained water
Natural lagooning	325	302
Macrophyte lagooning	325	297

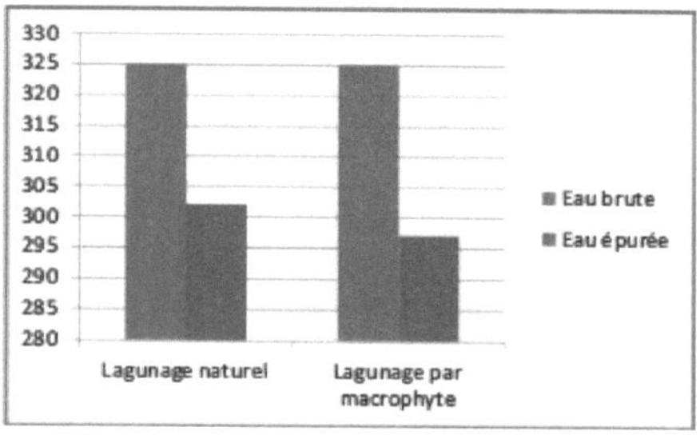

Figure 26 - Results of COD analyses.

Ammoniacal nitrogen NH4

Table 11- NH4 analyses

Samples taken	Raw water	Drained water
Natural lagooning	5.98	4.28
Macrophyte lagooning	5.98	3.66

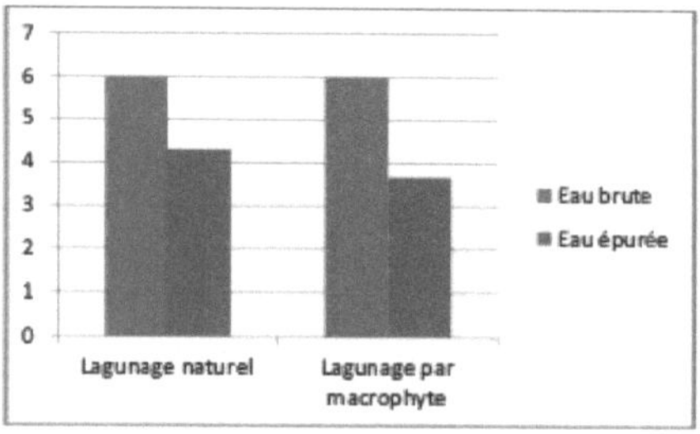

Figure 27 - NH4 analysis results.

The decrease in nitrogen concentrations in the experimental pilot plant means that the reeds are playing a role in absorbing NH4 thanks to their root system, which captures the organic load.

Conclusion:

It is clear from this literature review that phytodepuration offers a definite opportunity because of its multiple economic and ecological benefits.

The process consists of setting up a series of tanks, filled with gravel and pebbles, to support vegetation such as phragmites, typha, rushes, phalaris, etc., which use their powerful root system to extract mineral and organic pollutants.

This is an eco-biological process in which the wastewater is kept below the surface of the plant beds and circulates in different types of flow, vertical, horizontal or mixed, generating fewer odour nuisances.

Because of its simplicity, this depollution system could open up promising prospects for sustainable development, i.e. the creation of added value (use of treated water in agriculture, forestry, fishing, etc.) and the improvement of people's quality of life.

III.2 Brezina pilot plant for phytodepuration

Nowadays, the use of natural purification techniques is becoming increasingly popular for a variety of reasons, including technological, energy, economic and management considerations. Experiments around the world show the growing use of phyto-purification,

using plant beds of macrophyte vegetation.
This interest is all the greater on the outskirts of oasis towns, which are affected by insufficient water resources, with a view to protecting the natural environment from environmental damage and, if necessary, recovering and recycling treated waste water for the purpose of growing forest species to protect against silting and the advance of the desert.
This contribution deals with the technical aspects of installing a purification system based on the phytodepuration process. The experimental pilot plant, designed to treat domestic wastewater, covers an area of 0.3 ha. At the end of the chain, the waste water will be reused to irrigate a 5-hectare area of oasis trees, made up of various forage species adapted to the local agro-climatic context.

111.2.1 Description of the driver

The Brezina oasis is located in the Presaharan region (33°6'N, 1°15'E, 800 m s/m), 85 km south of the wilaya capital El Bayadh in Algeria. It lies on the syncline between the last ridge of the Atlas and the Saharan platform of the Erg Occidental. It is bounded by the Oued Seggueur, which drains a catchment area of 3,680 km^2 (Fig.28).

Figure 28- Location of the Brezina study area

The latter feeds the water table adjacent to the palm grove of the Brezina oasis, and guarantees the survival of the areas downstream from it, as far as the Daiet El Bagra depression. The construction of the Khang Larouia dam, at the confluence of the Ghasoul and Rahoul tributaries of the Seggueur wadi, has radically altered the hydrogeological balance of the Brezina oasis and the entire downstream region, making it necessary to make rational and optimum use of its conventional and non-conventional water resources.

111.2.2 Site characteristics

The natural context of the Brezina oasis is characterised by an arid Presaharan climate, with average temperatures ranging from a minimum of 9°C in January to a maximum of 35°C in August, with daily temperature variations in excess of 10°C. Precipitation varies between 100 and 180 mm per year; it is distributed throughout the year, with a higher concentration during the period from May to October. In summer it is sporadic and intense. Relative humidity is around 25% during the summer months.
The wind sweeps across the Brezina oasis from south to north, with average speeds in excess of 3 m/s. During sandstorms, the wind can exceed 45 km/hour. Solar radiation has average values of 4.59 Kwh.m^{-2} , relative to a coverage index that shows very limited seasonal variations, with an average value of 0.61. Evapotranspiration is around 1600 mm/year.

These factors give rise to high evapotranspiration of around 1600 mm per year, affecting the plant cover in the intervention zone, which is mainly composed of xerophilous species such as *Salsola vermiculata* and *Artrophytum schimittianum. Tamarix balacae* and *Peganum harmal* grow on the banks bordering Oued Seggeuer.
Surface water is mobilised by means of the *Kheneg Larouia* dam, a dam fed by a catchment area of 3680 km^2 , of the concrete-weight-road type with an overflow at 910 NGA, with a dike height of 60 m, creating a reservoir with a capacity of 122 Hm^3 , of which 62 Hm^3 represent the useful volume. However, given the very high rate of evaporation from the water surface, the useful volume of the reservoir means that only 11.5 Hm^3 /year can be regulated.
The characteristics of this catchment include low average annual flows (0.84 m^3 /s), a low runoff coefficient (2.88%), a low water level (7.21 mm), a low flow modulus (0,228 l/s/km2), a high abrasion rate (412 t/km2/year), rare but significant floods (882 m^3 /s), a high sedimentation rate (1.008 Mm^3 /year) and significant evaporation from the reservoir surface (2,500 mm/year).
The groundwater is located in the alluvial aquifer (Oued Seggueur), made up of heterogeneous sands and other permeable materials which lie continuously on an impermeable substratum. With a length of 16 km, the aquifer bypasses the village of Brezina to the south and continues for several hundred km before merging with the Tertiary deposits of the Great Western Erg.
The morphology of the local water table, which varies in width between 300 and 1500 m, gives the alluvium an average thickness of 10 m and an average storage coefficient of 7%, giving an estimated water reserve capacity of 8.96 Mm .^{3}
In economic terms, agro-pastoralism is the only source of income for the local population, representing a useful agricultural area of 1,249 ha, for a cadastral area of 13,180 km^2 , or 0.09%. The arid climate and the degradation of the vegetation cover due to overgrazing have led to a decline in the livestock population, which totals 249,728 head, made up of sheep, goats and camels.
The Brezina palm grove, which is home to the area's main economic activities, runs alongside the Oued Seggueur and is bordered by a strip of dunes formed by quaternary deposits, except for the southern part which spreads out over a marly hillside.
Covering an area of 174 ha, including 53 ha of productive seedlings, it contains 24,600 seedlings. In the cultivated areas, palm trees are grown in association with intercrops (Fig.29).

Figure 29 - Degradation and silting of the palm grove

111.2.3 Description of the Brezina experimental pilot

a) The principle :

The principle adopted for the Brezina pilot plant is based on the creation of two treatment processes, characterised as follows:

- **The "A" process:** This consists of a horizontal flow purification system designed to treat an average flow of 20 m^3 of water per day. The treatment process consists of three consecutive cells, organised as follows:
- *The first sedimentation cell is characterised by a water level that does not exceed 30 cm; this is where primary sedimentation processes and the first aerobic digestion take place. Vegetation outcrops and its density does not exceed 90%.*
- *The second cell carries out the anaerobic digestion phase, the water level is around 60 cm and the vegetation is sparse, occupying only around 20% of the available surface and concentrated near the entrance and exit of the cell.*
- *The third cell is used to complete the aerobic digestion of the wastewater, which is now in an advanced state of treatment. It is similar to the first cell, with a plant density of no more than 90% and a water level of less than 30 cm.*
- *The distribution system consists of PVC pipes (SN1110) laid at a gradient of 1-2%.*
- **The "B" system:** This is a horizontal surface flow system. It can treat 10 m3 of water per day, achieving a level of purification in line with the restricted irrigation limits indicated by the WHO. The system consists of three consecutive cells of different sizes filled with gravel of different sizes. In particular, the characteristics of each cell are as follows:
- *An inlet and outlet section where the inlet and drainage pipes are laid, filled with round gravel (60 -100 mm diameter).*
- *An internal section filled with round gravel with a diameter of 10 - 25 mm in which plant species are planted at regular intervals.*
- *Basins are waterproofed by compacting the sand that makes up the walls of the basins, laying a geotextile liner and a compacted layer of clay at least 20 cm thick. Alternatively, the walls can be sealed with a high-density polyethylene lining at least 1mm thick.*
- *The wastewater distribution system is gravity fed. The PVC pipes (SN1110) are laid at an incline of 1-2%.*

111.2.4 Architecture of the experimental pilot

The morphology and topography of the land limit the size of the installation and require the treatment of a fraction (30%) of the wastewater flow discharged by the agglomeration. The treatment process is equipped with a bypass, a 0200 mm inspection manhole and a 1% slope to ensure that the water flows by gravity towards the treatment plants.

A primary sedimentation tank (*imhoff* type*)* and a grading unit are used to remove solids. Finally, there is a valve to regulate and measure the flow to each of the planned treatment plants. It should be noted that the population of the Brezina agglomeration is estimated at 11,800 (RGPG 2008), representing a wastewater discharge of 9 l/s (780 m3/d).

The two wastewater treatment plants discharge the treated wastewater into a common accumulation tank with a dual function: to provide an additional cycle of sedimentation and anaerobic microbial digestion, and to store water to guarantee the arboretum's supply.

Its dimensions are a depth of 2 m and a slope of 2/1. The water communication between the various pools consists of PVC pipes laid at an incline of 1 to 2%, positioned symmetrically in relation to the longitudinal axis of each pool. The pipe inlets and outlets are protected by a grid to prevent small animals from entering. Access is provided to all water distribution pipe

connection points to ensure proper maintenance of the installation.
The basins are sealed with a compacted layer of clay at least 20 cm thick and a high-density polyethylene (HDPE, 1.5 mm) lining, including on the basin walls to seal off run-off. The banks are covered with a layer of fine gravel with a diameter of 10-50 mm, to protect them from water erosion, especially in the water inlet section (Fig.30/31).

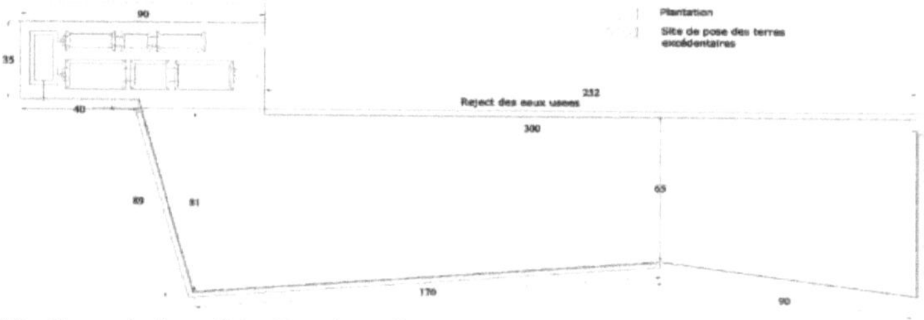

Figure 30- Ground plan of the Brezina pilot plant for phytodepuration

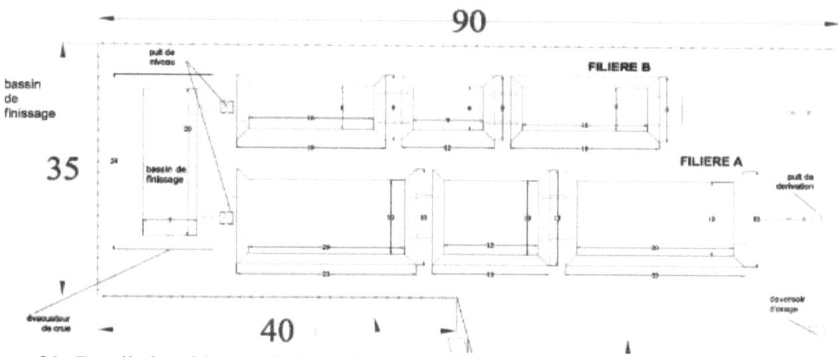

Figure 31- Detailed architectural plan of the phytodepuration plant

111.2.5 Standards for the use of TMEs in irrigation

The purification objectives for which the plant was designed are to bring the current quality of wastewater discharges up to discharge standards, in compliance with national standards (Executive Decree no. 93-160 of 10 July 1993) and those accepted by the World Health Organisation (WHO, 1989) (Table 12).

Table 12- Characteristics of the pollutant load of Brezina wastewater

Parameter	Parameters	Unit	Current quality	Standards (WHO)
Suspended matter	MY	mg/l	417	35
Biological demand in	BOD5	mg/l	161	25
Total phosphorus	TP	mg/l	0.95	2
Total nitrogen	TN	mg/l	49	30
Pathogen	FC	cfu/100 ml	100,000	1,000
nematode eggs	N	cfu/100 ml	0,1	0.1

The aim of such a process is to obtain water that meets conventional standards and can be used for irrigation, in particular with two bacteriological limits, in line with WHO (World Health Organisation) recommendations.

In Algeria, the regulations on the use of waste water prohibit all types of use of unpurified water for agricultural purposes, particularly for raw consumption. The latter requires tertiary treatment (sterilisation, chlorination): a maximum of 30 mg/l for BOD_5 , the elimination of pathogenic E. coli agents to less than 1,000 per 100 ml of water and 0.1 eggs of nematodes per 100 ml of water.
However, the appropriate use of purified water is authorised for certain crops for industrial use, forestry and ornamental green spaces and dune fixation, fodder for livestock feed, etc. (Fig.32).

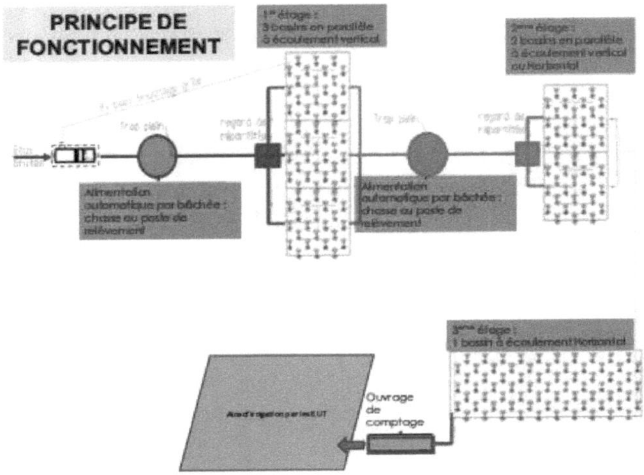

Figure 32- Operating principle of a macrophyte WWTP

CONCLUSION

From the parameters listed above, it is clear that the construction of a treatment plant using the phytotreatment process is essential and entirely justifiable if this type of wastewater treatment is to be extended to all the country's small and medium-sized urban communities. In addition, this phyto-purification alternative offers a better opportunity, in particular through the rational use of non-conventional water resources, while ensuring competitive levels of purification performance at moderate cost.

Monitoring of the purification performance of this experimental pilot in an arid region (Brezina Oasis, in the wilaya of El Bayadh) in partnership with the University of Vertigo (Italy), has revealed significant indicators of the role of the combination of the plant-sand-solar energy complex in the elimination of undesirable and polluting matter, in both physico-chemical and bacteriological terms.

Through this project, which comes under the CDM (Clean Development Mechanism) initiated by the Kyoto Convention and applied to an oasis area where the availability of water is the main limiting factor for development, the results of reducing wastewater pollution using simple, ecological processes will help to meet the objectives set out in the new water strategy in Algeria since 2005.

In fact, the implementation of a natural water purification pilot scheme at Brezina would provide a sufficient quantity of water to allow forestry plants to grow over large areas, which would ultimately lead to sustainable benefits such as reduced wind erosion, improved soil fertility and a general increase in ecosystem biodiversity.

In economic terms, the diversification of fodder, forestry and livestock production, as well as the provision of employment, biomass energy, the production of carbon emission credits, and other secondary products such as income from pharmacological and cosmetic uses (e.g. vegetable oils), etc., will be of great benefit.

It is important to stress that such a project has already contributed to triggering a fruitful debate on the issues of environmental sanitation linked to the convention on climate change and the role of research for the development of arid zones and degraded ecosystems. The project also provides an opportunity to draw the attention of the public authorities responsible for cooperation and sustainable development in Algeria.

When we set out to purify water, whether used or spring water, the water purification technique or process multiplies and diversifies, all the while leading to a less polluting purification process that allows appropriate use of the resultant water.

After a water purification process, we can end up with water that is completely drinkable, or water for agricultural use, or water for household use. And all this depends above all on the technique adopted, which requires the right tools and the right means to achieve it.

In Algeria, wastewater treatment plants do not meet the drinking water standard, given the techniques and resources adopted. Natural lagooning is the technique adopted by current wastewater treatment plants, resulting in water for agricultural use.

In the study we carried out, we proposed another wastewater treatment technique, macrophyte lagooning. This process still falls within the range of techniques used to treat water for agricultural and domestic use, but the results are more promising.

It should be stressed that the potability or usability of water depends on the purity of the water, which represents the level of physico-chemical and biological parameters in the water.

In fact, analysis of the values of the physico-chemical parameters after implementation of macrophyte lagoon treatment shows that this technique can produce healthier and purer water than that produced by the natural lagoon technique, given the lower level of physico-chemical parameters in the water, which leads us, at the end of this study, to confirm our hypothesis proposed in the introduction.

Unfortunately, this macrophyte lagooning technique has not been adopted in Algerian wastewater treatment plants, and the simplicity of the means and materials it requires, and above all the results it can achieve, make it a very attractive option for the development of biomass.

We call on the relevant authorities to take full account of this macrophyte lagooning technique and to adopt it in the wastewater treatment plants currently under construction, so as to achieve good quality wastewater.

REFERENCES

[1] Mettahri, M.S. (2012). Simultaneous elimination of nitrogen and phosphate pollution from treated wastewater using mixed precedents. Cas de la STEP de la ville de Tizi-Ouzou. [These de Doctorat]: agronomie, genie des processes. [Tizi Ouzou]: Universite de mouloud Memmeri. 4; 21; 23; 24.

[2] Rodier, J. (2005). L'analyse de l'eau naturel, eau residuaire et l'eau de mer. Paris: Dunod.

[3] Crosclaude G. (1999). L'eau, toml : milieu naturel et maitrise et tome 2 : usage et polluants Versailles, INRA, (Paris).

[4] Maji solution 2020nh : Treatment of sewage sludge and used water, presentation leaflets, Water and sewerage glossary, Zerif Lite Developed by Themeles, 2020.

[5] Mathieu, C. and Pieltain, F. (2003) : Analyse chimique des sols methodes choisies. Editions Tec et Doc/Lavoisier, Paris.

[6] Abibsi Nadjet 2011: Reuse of wastewater through plant filters (phytodepuration) for irrigation of green spaces - application to a district of the city of Biskra

[7] Rejsek, F. (2002). Analyse des eaux ; aspecte reglementaires et technique. Ed Canope CRDP de Bordeaux (France).

[8] Hamadou Hadjer 2019: Analyse physico-chimique et bacteriologique des eaux uses de STEP Boumerdes (W. Boumerdes), Rapport de soutenance.

[9] Elmund et al (1999). Comparison of Escherichia coli, total coliform and fecal coliform populations as indicators of wastewater treatment efficiency. Water Environ. Res. 332-339.

[10] Barthe C., Perron J. et. Perron J.M.R. (1998). Guide d'interprétation des parametres microbiologiques d'interet dans le domaine de l'eau potable. Quebec.

[11] WHO (2000). Guidelines for drinking-water quality; volume 2; hygiene criteria. World Health Organization, 2nd edition.

[12] Swiss Medical Journal 2014 :

[13] Mentegut, 1987. The aquatic environment and flora. Tome 1. Association de coordination technique agricole (ACTA) edition, ISBN-2-85794-061-0 Paris.

[14] Pronost J; Pronost R; Deplat L; Malrieu J and Berland J.M. 2002. Station d'epuration, dispositions constructives pour ameliorer leur fonctionnement et faciliter leur exploitation, documents techniques FNDAE N°22 bis, PDF, Page 73.

[15] Menard J.L, Coillard J., Lienard A., Chabenat A., 2005. Les influents peuvent charges en elevage du ruminons, procede de gestion et traitement valides pour une mise en conformite plus econome, collection synthese.

[16] Kirere Mathe, 2001. Cours d'assainissement, de premiere graduation, section sante communautaire. Institut superieur des techniques medicales, Nyon Kunde.

[17] Dabouineau L., Yann, Philippe Collas, 2005. Phytoremediation or phytoremediation or the use of plants for the depollution and purification of waste water.
U.C.O. Bretagne Nord a Guingamp, le role d'eau vol 124.

[18] Meliani H.A, 2006. Role of vegetation in the semi-arid of north-western Algeria. Mem. Magister. Mascara.

[19] Dornano M. and Mehignerie P., 1983. Lagunage naturel et lagunage aere procede d'épuration des petites collectives.3eme edition. Tours. France.

[20] Pascal Molle, 2003. Filtres plantees de roseaux, limite hydraulique et retention du phosphore, Université du Montpellier 2, science et technique du Languedoc. These.

[21] Jund S., Paillard C., Andre P., Frossard B. C., 2000: Guide de gestion de la végétation des bords de cours d'eau, Rapport generale agence de l'eau Rhin-Meuse.

Annex 1: Calculation of the dimensions and estimate of a plant for phytodepuration (Case of the community of El KEURT-2000 E.H)

Appendix 2: Recommended water plants

Appendix *1:* Calculation of the dimensions and estimated cost of a plant for phytodepuration (Case of the community of El KEURT - 2000 E.H)

I./ PROJECT BACKGROUND

The El Keurt lagooning plant, which is the subject of this contribution, is administratively part of the commune of El Keurt (wilaya of Mascara). Geographically, it is located in north-west Algeria, between 35° 07' and 35° 31' north latitude and between 0° 0' and 0° 26' east longitude. Hydrographically, the study site is part of the Oued Maoussa catchment area, which flows into the Oued El Hammam.

The study area covers 10 ha of flat land at an average altitude of 470 m. Hydrogeologically, the plain contains a groundwater table with high mobilisable reserves, estimated at 61 Hm^3 (Sourisseau, 1974). It is bounded to the north by the RN7 trunk road and to the west, east and south by the agricultural land of the Mascara plain (Fig. 33).

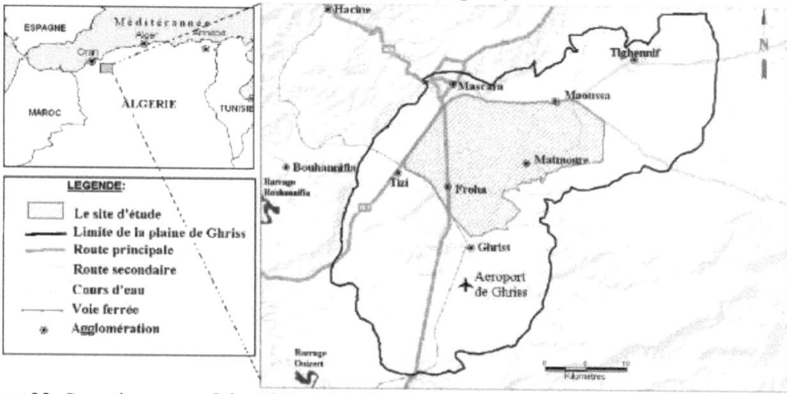

Figure 33- Location map of the El Keurt study area.

- **Population**: Out of a municipal population of 4,500, the conurbation has a population of 2,300 (2021 estimate).
- **Social facilities** :
- A primary school and a secondary school with a total of 714 pupils.
- A treatment room, a rural maternity hospital ;
- Hammam ;
- A play area,
- A mosque
- A postal agency.
- **Drinking water supply networks :**
- Allocation: 120 l/d/capita.
- 01 Drilling 250 ml deep (Ghriss aquifer).
- Flow rate: 7 l/s.
- Delivery station.
- Reservoir and distribution network with individual connections.
- Distribution: 1 day out of 2;
- The water supply network is managed by ADE (Algérie des Eaux).

- Estimated real allocated flow: Qe.p = [(120 x 1800) x 0.7)] / 86400 = 1.75 l/s.
- **Sewerage system and wastewater discharge area :**
- Combined sewer system.
- 2 km drainage collector.
- Discharge points: Natural lagooning plant.
- The discharge point, fenced off by the zimarman, has a surface area of 1,200 m^2 or 30 x 50 m.
- Management of the sewerage system by the local authority.
- Estimated flow to be treated: Q e.u = Q e.p x 0.6 = 1.05 l/s.

- **Land**: The land adjoins the RN17A, a state-owned property.
- **Soil and slope**: very good soil for gravity drainage.

Sandy-silt formation. Downstream, an agricultural activity irrigated by wells (30 m deep). At a recent date, the crops, irrigated by used water, were confiscated and incinerated because of contamination by pathogenic germs, a source of water-borne diseases.

- **Main activity**: mountain farming :

- Cereal growing: 107 Ha ;
- Viticulture: 63 Ha ;
- Arboriculture: 13 Ha ;

- **Description of the existing lagooning plant**

The *El* Keurta lagoon treatment plant was built in 2010 as part of the sectoral water works programme. It is designed to treat domestic wastewater from the El Keurt agglomeration. Access to the plant is via the RN17A, linking Mascara to Bouhanifia (Figs. 34, 35 and 36).

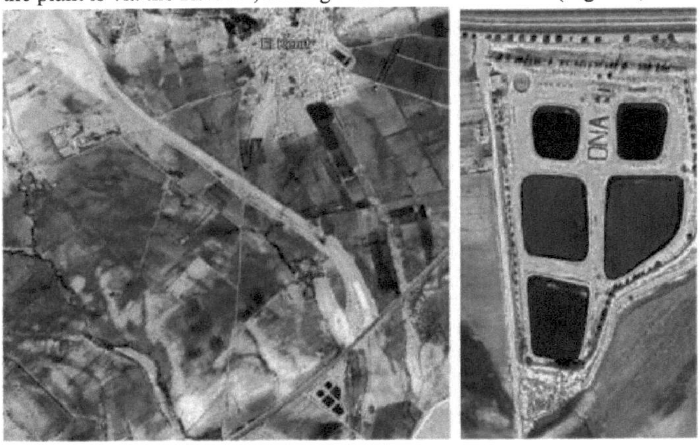

Figure 34 - Location of the El Keurt lagoon

- **Description of installations**

The El Keurt wastewater treatment plant uses a natural lagooning process.

- **Purification basin** (anaerobic lagoons)

-Water flow to be treated: Q-1500m3/d

Purification constant at 20°c: K-0.10^1 BOD, influent: 246mg/L

-Temperature coefficient (constant): Z-1.06 Feed water temperature in winter is 15°c -Stay time at 20°c T, -20 days

-Length L1= L2=125m, Width 11-12=40m, Height H=3 mL

Figure 35 - Pond 01 (anaerobic). Figure 36 - Basin 02 (anaerobic).

-Optional pools

Aerobic lagoons (for maturation):

- Water flow to be treated: Q-1500m/
- Feed water temperature in winter is 14°c
- Stays at 20°C T, -7 days
- Length 11-12-115m,
- Width 11-12-38m,
- Height H= 1.2 m

II./ SIZING CALCULATIONS

11.1Site topography

a) Desired discharge level: In Algeria, discharge standards are set out in implementing decree no. 93-160 of 10 July 1993, which lays down maximum limit values for discharge parameters.

b) Number of equivalent inhabitants (EH): For the municipality of El Keurt, we consider that the number of ***equivalent inhabitants*** *(*EH) is 2,000 EH, given its current population (estimated by the APC) to which must be added the school flows from the rural area that the schools in the main town receive on a daily basis.

c) Denivelee: This is the hydraulic gradient i, corresponding to the slope of the water line in the gravel pack between the inlet and the outlet. The nature of the terrain separating the built-up area from the plant site is of the order of 3%**.**

11.2Determining the number of storeys

Remembering that the determination of the number of stages and the type of flow (V and H), we have opted for a classic diagram, which has given good results elsewhere, in a Mediterranean context.

Through this pilot station in El KEURT, we intend to study its operation over several years with a view to adapting it better to the local and regional context. A three (3) storey station has been selected.

III. CALCULATION NOTE

d) Debit :

The flow rate was determined as follows:

- Average flow to be treated per inhabitant : $Q_m = 0.120 \times 0.7 = 0.084$ m^3 /d/EH);
- the average daily flow to be treated: $Q_T = Q_m$ * p.e. = 0.084 x 1,800 = **151.2 m^3** /d;
- instantaneous flow :Q! = (Q_T * 103) / 86400 = **1.75 l/s** ;
- peak flow : Qp= kc * Qi. Where Kc = 3; Qp = **5.25 l/s**

b) Determining the number of storeys

Remembering that the determination of the number of stages and the type of flow (V and H), we have opted for a classic scheme, which has given good results elsewhere, in a Mediterranean context. Through this pilot station at El KEURT, we intend to study its operation over several years with a view to adapting it better to the local and regional context. As a reminder, the expected performances of this system are as follows:

- Vertical floor:

- MES :80% ;
- BOD5 :55% ;
- COD :55% ;
- N-NK :30% ;
- PT:10%.

- Horizontal floor:

- MES :20% ;
- BOD5 :50% ;
- COD :50% ;
- N-NK : 20% ;
- PT:50%.

A three (3) storey station has therefore been selected:

- A horizontal first stage consisting of 3 basins, enabling most of the TSS and almost half of the BOD5 and COD to be removed, along with the beginnings of denitrification; planted with phragmites.
- A vertical second stage to complete the abatement of the first parameters and begin the elimination of phosphates; phragmites + rush plant.
- A third vertical level for polishing the water, consisting of 2 basins separated by a passageway, planted with different types of macrophytes (phragmites, rushes, cattails, bulrushes, etc.).

Given the number of stages, the choice of flow (Horizontal or Vertical) depending on the performance required and whether or not recirculation is required, it is possible to calculate the surface area of each stage.

By empirical convention, you should know that in one day, the height of water that can pass through a pool on the 1er floor is *40 cm*.

Hence :

$$Sba\ 1^{er}\ floor = Q / 0.4 = 151.2/0.4 = 378\ m^2$$

So the surface area for the whole of the 1er floor is multiplied (x) by the number of pools.

For three (3) basins the surface area is :

$S1^{er}$ floor = Sba 1er * 3 = 378 x 3 = 1 134 m2

- On the 2eme floor, the height of water that can flow into a basin in one day is ***One (1) metre***, as the effluent has already been filtered. Hence the surface area for the 2eme floor: Sba 2eme floor = Q/1 = 151.2 m2

The surface area of the 2nd floor is therefore doubled because there are two (2) pools:

$S2^{eme}$ floor = Sba 2eme floor * 2 = 151.2 x 2 = 302.4 m2

- In the case of the top floor (Horizontal), it should be borne in mind that an inhabitant needs *2.8 m^2* to achieve proper purification. To calculate the surface area of the 3eme floor, we therefore need to calculate the total surface area (Stot):

$$Stot = 2.8* E.H = 2.8 \times 1,800 = 5,040 \text{ m2}$$

Hence the surface area of the 3eme floor:

S3eme floor = Stot - (S1er floor + S2eme floor) = 5 040 - (1 134 + 302.4) = 3 603.6 m2

Once the total surface area of the basins is known, we need to add the surface area of structures such as the screen, manholes, fencing and various easements, etc., which leads us to a mark-up rate of 100%, multiplied by 2. Hence the overall surface area of the station:

Total surface area of the station = Stot *2 = 5 040 x 2 = 10 080 m2, i.e. approximately 1 Ha

Having obtained the surface area of the entire plant, all that remains to be done is to draw up a plan to scale using a topographical survey, in order to be able to lay out the floors, basins, bachees tanks, degrilleur,...etc.

A soil survey will be carried out (hand auger) and laboratory analyses of several soil samples will be carried out to determine the dominant structure and the measures to be implemented to ensure that the site is watertight in accordance with regulations. A geological and hydrogeological analysis of the site will be useful to support the measures to be considered.

Number of hatches per floor :

On the vertical levels, the water is sent in small quantities, but with a high load so that it can be distributed over the entire surface area of the basin; this is known as *beaker* delivery.

On the 1er floor, the desired water level is *2 cm* and on the 2eme floor it is *3 cm*. The latter, which is slightly higher because the passage on the 1er floor allows the effluent to infiltrate, will not be useful for our plant.

It is necessary to know the number of covers per storey in order to determine the volume of the structure and its construction cost:

- Number of tarpaulins on the 1st floor :

Bl = blade * Sba 1er floor

Bl = 0.02 * 1134

Bl= 23 caches

On the basis of these data, the sizing of the macrophyte WWTP in the municipality of El Keurt can be established as follows:

IV./PROJECT ESTIMATE

The reference year for this estimate for the project to build a wastewater treatment plant with an average capacity of between 2000 Eq.Hab is 2022.

The costs of the design and monitoring study are not included, as this assumes that these services are the responsibility of the research team, attached to the university laboratory.

This estimate has been drawn up on an indicative basis, exclusive of tax, to give local authorities an overall idea of the financial resources to be mobilised.

Table 12 - Design and project management costs

Designation	Unit	Quantity	P.U (in DA)	Amount before tax (in DA)
1. Topographic survey	U	1	200 000,00	100 000,00
2. Geotechnical study	U	1	200 000,00	100 000,00
3. APS-APD	u	1	1 500 000,00	600 000,00
4. Laboratory costs	и	1	200 000,00	200 000,00
5. Technical assistance	и	1	1 200 000	1 000 000,00
TOTAL (2)	-	-	-	**2 000 000,00**

Table 13 - Cost of construction work

DESIGNATION	Unit	Quantity	P.U (DA)	Amount (DA)
1. Installation and earthworks:				
- Stripping topsoil	m^3	4 480,0	50,00	224 000,00
- Trench/channel excavations	ml	51,0	250,00	12 750,00
- Digging the basins	m^3	16 800,0	100,00	1 680 000,00
			S/TOTAL	**1 916 750,00**
2. Piping :				
- PVC-0 90 pipes	ml	210,5	450,00	94 725,00
- PVC pipes - 0 120	ml	101,9	650,00	66 235,00
- Supports	U	36	250,00	9 000,00
			S/TOTAL	**169 960,00**
3. Structures and masonry :				
- Manual tumbler	U	1	2 500,00	2 500,00
- Automatic flush - Etagel	U	1	15 000,00	15 000,00
- Regards	U	5	1 500,00	7 500,00
			S/TOTAL	**25 000,00**
4. Lining and seals :				
- Geotextile anti-poin^onnement....	m^2	6 687,8	150,00	1 003 170,00
- 1 mm PE geomembrane	m^2	6 687,8	250,00	1 671 950,00
- Draining materials, pebbles 30/60...	m^3	426,0	300,00	127 800,00
- Filter materials, gravel 4/12...	m^3	2 824,0	300,00	847 200,00
- Road drains	ml	143,4	350,00	50 190,00
- Agricultural aeration drains	ml	916,8	150,00	137 520,00
- Ventilation ducts	ml	10,0	750,00	7 500,00
- Ventilation caps	U	10	250,00	2 500,00
- Energy dissipation plates...	U	24	100,00	2 400,00
- Locker divider plates.	ml	44,7	300,00	13 410,00
			S/TOTAL	**3 863 640,00**
5. Miscellaneous equipment:				
- Closing	ml	736,0	500,00	368 000,00
- Access door	U	1	7 500,00	7 500,00
- Lodge - shop	U	1	120 000,00	120 000,00
- Installation of reeds	m^3	20 160	25,00	504 000,00
- Replacing topsoil....	m^2	3 488,0	150,00	523 200,00
			S/TOTAL	**1 522 700,00**
TOTAL	-	-	-	**7 498 050,00**

The total cost of a plant with a capacity of 2,000 population equivalents is estimated at around 7,498,050.00 DZ-HT (excluding tax), i.e. a specific cost of 4,165 DZ / population equivalent. By way of illustration, a natural lagooning plant, such as that at El Keurt (the reference plant), is valued at ten (10) times the cost. Added to this are operating costs (3 permanent jobs) and maintenance costs corresponding to the periodic cleaning of the sludge.

Figure 37- Diagram of a phytodepuration plant

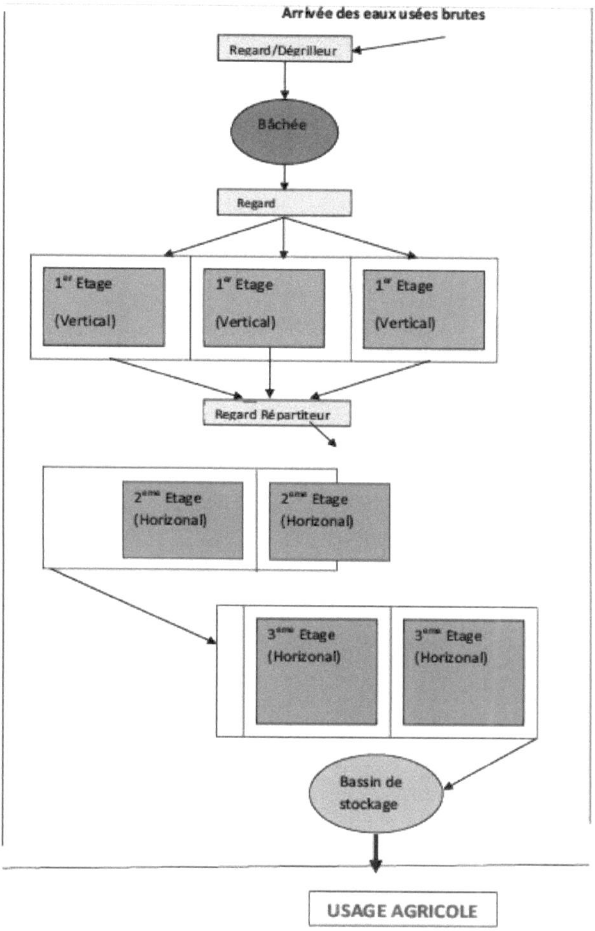
Arrivée des eaux usées brutes
Regard/Dégrilleur
Bâchée
Regard
1er Etage
(Vertical)
1er Etage
(Vertical)
1er Etage
(Vertical)
Regard Répartiteur
2ème Etage
(Horizonal)
2ème Etage
(Horizonal)
3ème Etage
(Horizonal)
3ème Etage
(Horizonal)
Bassin de
stockage
USAGE AGRICOLE

Appendix 2:

Recommended water plants

Helophytic type (roots in the water, aerial leaves). These are large herbaceous shore plants such as :
Common reed;
*Tupha latifolia (*Broad-leaved sledgehammer);
*Joncussp. (*Jones);
Lake bulrush (scirpus lacustris);
Sparganiumsp. (ribbon tree).
Alisma plantago (water plantain);
Alopercurus geniculatus (vulpine grass);
Carexsp. (Laiches);
Fireweed (*Epilobium sp.*);
Pseudacours Iris (Water Iris);
Lycopus europarus (European Lycopus);
Mentha aguatica (water mint);
Myosotispalastris (marsh forget-me-not);
Phalaris arundinacera (false reed, baldingere).

Figure 38 - Typha latifolia

Plants of the edges of ponds forming a belt, which replaces the Roseliere on the somewhat steep banks.

Figure 39 - Scripus palustris

This plant is mainly found on the edges of ponds; it only grows on fairly flat, sandy-clay banks. Can also be found in marshes and in all damp places, but with a constant surface water eutrophic water.

Figure 40 - Juncus sp. This species is typical of marecages and poorly drained grasslands, and is highly resistant to summer drought.

Figure 41 - Arundo plinii

It is one of the most widespread aquatic grasses in the world, highly resistant to drought in low-water periods. However, it is mainly in the pond belt that it is most dynamic.

dynamic.

Poaceae: a riparian species, growing mainly at the foot of solid banks, or on muddy banks, which can venture into marshes but is confined to low, slightly flooded banks.

Figure 42 - Cyperus langus Figure 43 - Cyperus esculentus

Figure 44 - Rumex sp.

A gramineous species typical of watersides, either on consolidated banks or on the mud of ponds.

Figure 45 - ***Phragmites communis,*** *a* species used to treat waste water

MIX
Papier aus verantwortungsvollen Quellen
Paper from responsible sources
FSC® C105338

Printed by Books on Demand GmbH, Norderstedt / Germany